MW00851880

# Praise for *Reentry*

"I have followed the rise of SpaceX for almost two decades, and flown on the company's rockets and spacecraft. But only after reading *Reentry* can I appreciate the challenges SpaceX overcame to launch a new space age."

**—Michael López-Alegría, six-time astronaut**

"Eric's books, *Liftoff* and now *Reentry*, capture the story of perhaps the world's most remarkable and ambitious corporate endeavor, led by an exceptionally bold and accomplished visionary in Elon Musk. These works take the reader on a behind-the-scenes journey through SpaceX's highs and lows, the flaws, failures, and world-changing accomplishments."

**—Jared Isaacman, commander of the first all-civilian spaceflight**

"Ever wonder how Elon Musk and SpaceX have managed to single-handedly disrupt the entrenched and all-powerful aerospace industry in record time? Eric Berger takes us inside the meeting rooms, factories, and test stands where it all happened. Colored with personalities, technical details, and context, *Reentry* tells the as-yet-untold story of the drive and determination it took to reuse rockets and launch a new space age."

**—Lori Garver, former NASA deputy administrator and author of *Escaping Gravity***

"In *Reentry*, Eric Berger goes behind the scenes to tell the stories of the heroes behind SpaceX's success. He lays bare what it takes to succeed in an unforgiving business, while also leaving the reader questioning if there might be another way. The result is a brilliant narrative of individual and collective toil, strife, sacrifice, and ultimate triumph—a must-read, and nothing less than a critical addition to humanity's historical record."

**—Andy Lapsa, cofounder and CEO of Stoke Space**

**Also by Eric Berger**

*Liftoff: Elon Musk and the Desperate Early Days that Launched SpaceX*

# REENTRY

## SPACEX, ELON MUSK, AND THE REUSABLE ROCKETS THAT LAUNCHED A SECOND SPACE AGE

ERIC BERGER

BenBella Books, Inc.
Dallas, TX

BenBella Books, Inc.
10440 N. Central Expressway
Suite 800
Dallas, TX 75231
benbellabooks.com
Send feedback to feedback@benbellabooks.com

*BenBella* is a federally registered trademark.

Printed in the United States of America
10 9 8 7 6 5 4 3 2 1

Library of Congress Control Number: 2024007021
ISBN 9781637745274 (hardcover)
ISBN 9781637745281 (electronic)

Editing by Rick Chillot
Copyediting by Elizabeth Degenhard
Proofreading by Becky Maines and Jenny Bridges
Indexing by WordCo Indexing Services, Inc.
Text design and composition by PerfecType, Nashville, TN
Cover design by Sarah Avinger
Cover photography by John Kraus
Printed by Lake Book Manufacturing

*For the thousands of people at SpaceX who made the magic happen.*

# CONTENTS

# PROLOGUE

*April 20, 2023*
South Padre Island, Texas

Brown and barren scrublands tumble endlessly across southern Texas, undulating gently all the way down to the roily Rio Grande River. This land of heat and haze spreads to featureless horizons, a desolate expanse mostly unmarred by human hands. Yet on a sublime morning in the spring of 2023 a living, breathing monster rose above the hardscrabble grounds, tall as a skyscraper, with a shiny steel carapace glinting in the pale sunlight.

The largest and most powerful rocket ever built was puffing into life, ready to burst into the sky above and noisily proclaim a new era of spaceflight. This was Starship. After two decades of relentless toil, Elon Musk had pushed and cajoled and bullied and single-mindedly driven SpaceX to the precipice of history. His company had launched hundreds of rockets. But Starship was its first that might, one day, make good on Musk's fever dream of sending humans to Mars and other new worlds.

As Starship shimmered in the Texas sunshine, Musk shivered with nerves. His stress spikes around major launches, when the world watches and judges. "My gut is twisted right now due to the impending launch," he admitted to me shortly before Starship lifted off.

And why not? What stood on the launch pad represented both an epic achievement and a monumental risk. The world had never before seen its like. Decades earlier, with 5 percent of the U.S. budget at its disposal, NASA built the mighty Saturn V rocket and sent astronauts to the Moon. But Starship is a good deal larger, and far more ambitious. When the Saturn V ventured to the Moon more than fifty years ago it jettisoned its stages along the way, into the ocean and empty space. Only the cramped Apollo crew compartment came home, and these capsules went into museums after each flight.

With Starship, Musk and his company SpaceX sought to build an entirely reusable rocket. The first stage of the rocket, named Super Heavy, would fly back to the launch site and hover, to be plucked from the air by large robotic arms and refueled for another flight soon after. The Starship upper stage would go onward into orbit, and from there to the Moon, Mars, or wherever. When its crew was ready to come home, they would simply fly back through Earth's atmosphere and make a powered landing where they pleased. Starship would then fly again and again.

That was the vision, at least. This first Starship, which triggered Musk's anxiety, was merely a full-sized prototype. Things could go wrong. Things probably would. SpaceX had invested billions of dollars into the launch facility, building a massive spaceport in a remote corner of the country. If the rocket exploded, the launch site and years of work were at risk.

The first seconds after liftoff were excruciating. The rocket's thirty-three engines sandblasted the launch site and surrounding wetlands, obscuring the view. A few engines failed almost immediately, in part due to flying debris from the launch pad, leaving Starship with barely enough thrust to start climbing. But eventually, second after second, climb it did. After clearing the massive dust storm, Starship started to pull away from the ground.

It was this colossal, almost impossibly huge thing. And there it was, long and silver and beautiful, furiously flying into the Texas sky. Musk could finally breathe a sigh of relief. The rocket flew more or less on course for about a minute and a half, far out over the Gulf of Mexico. But then, as problems with its multitude of engines increased, Starship started to wobble, and this triggered its self-destruct mechanism. As big things often do, the Starship era began with a bang.

A few days later Musk expressed appreciation for his employees. "I thought the SpaceX team did amazing work," he said. "And this is really one of the hardest technical projects that humanity's ever done. A fully reusable, humongous rocket. This is certainly a candidate for the hardest technical problem done by humans."

So how did they do it? How did a private company come to build the world's largest rocket? How did these engineers restore NASA's capability to fly astronauts into orbit, and then declare their intent to leap past the venerable space agency and send humans to Mars? How is SpaceX able to launch and operate more satellites than any other company or country in the world, by a factor of ten?

In *Reentry*, we'll dive into the great adventure of SpaceX: the hardships, setbacks, and ultimately the triumphs that led to the company becoming the most important actor in spaceflight today. By launching and landing their Falcon 9 rocket, the company has almost single-handedly spurred a massive sea change in the course of space exploration. To accomplish that, thousands of people sacrificed much in the last decade and a half, when SpaceX went from being incapable of launching a single rocket to putting nearly one hundred into space in a single year.

I've spoken with many dozens of SpaceX employees about what they did and how they did it. To know where SpaceX is going, and why they just might get there, it is critical to understand how its people built the future. This is their story.

It is also the story of Elon Musk and his outsized spaceflight ambitions. Moments after Starship broke apart in the sky, a conflagration of another sort broke out online. His critics seized on the arresting visual of an exploding rocket as yet another sign of Musk's incompetence, alongside the financial collapse of Twitter and various Tesla woes. Musk was a fraud, had always been a fraud, and finally the world was finding out. But with equal energy, defenders pointed out that Starship was an experimental rocket, and that SpaceX was willing to fail in order to go fast.

So what is the truth?

It is this: Musk founded SpaceX, has led it for two decades, and remains the principal visionary. He has made mistakes, and this book explores and explains them. Yet he also made the most consequential decisions, ones that have turned SpaceX into the innovative space powerhouse it has become. More often than not his instincts have been sound. On balance, then, SpaceX has been successful because of its mercurial leader, not in spite of him.

Yet nothing lasts forever. Musk is increasingly one of the most controversial and polarizing figures on the world stage. Where he once straddled the political divide with a strong Libertarian streak, Musk has increasingly sided with right-wing politics and alienated officials on whom SpaceX relies for government contracts. Some days, it seems like he is making political enemies as fast as SpaceX makes rockets. And with his business empire Musk has become deeply enmeshed in global conflicts, with interests in China, Ukraine, and elsewhere around the world that sometimes do not align with those of the United States government.

This may eventually force a reckoning. At present, SpaceX remains an absolutely critical contractor for both the American military and its allies, as well as NASA. The company stands, credibly, on the cusp of opening the way to Mars and beyond with a truly revolutionary launch system. But the world turns and inevitability fades. The biggest players sometimes make the biggest falls when confronted by fresh-faced upstarts.

Essentially, that is the very story told here.

|1|

# A VIOLENT BEAST

*November 22, 2008*
McGregor, Texas

Tom Mueller bounded up timeworn concrete steps, two at a time, into the vast and starless Texas night. After toiling so long on a spectacular new space machine, the wiry rocket scientist hastened to not miss its birth, with all the attendant sound and fury.

Less than half a mile away, a mighty rocket roared, its nine engines rapidly consuming half a million pounds of fuel. As these Merlin engines burned, they produced more than enough power to electrify the entire Los Angeles area—with every last one of its bright Hollywood lights—twice over.

During the countdown to this key test, Mueller had watched his computer monitors from the control room inside a concrete bunker. For more than a year he and his propulsion team at SpaceX had sweated the details of arranging nine engines, each a separate inferno, so they would burn brightly but not scorch one another. They had devised the intricate plumbing to deliver all of that rocket fuel. And they had managed the careful process of igniting the nine engines nearly simultaneously. Mueller's team had then clamped the rocket atop a large concrete tower, known simply as the "tripod," intending to test-fire it for 178 seconds as if the booster were launching into space.

The countdown had proceeded smoothly, and as each engine lit the computers reported no anomalies. All well and good. But as the rocket started to burn, Mueller wanted to *feel* the power of his creation, not just stare at numbers on a screen. Though the half-buried blockhouse lay close to the tripod, its sturdy walls muffled the test's sounds. "I'm going to watch!" Mueller shouted to the team of engineers monitoring the test at a clump of desks and monitors, running out of the bunker toward a roaring wall of brilliant light.

"There was a bright orange glow everywhere," he said. "And it was louder than hell."

Though the Falcon 9 era dawned brightly in Central Texas that night, its tremors barely registered in the broader aerospace industry. No one, really, took SpaceX all that seriously. In a little more than six years since its founding by Musk, the company had launched its small Falcon 1 rocket four times, and failed on three of those attempts. To date, the company had not lifted a single customer payload into orbit.

Despite this shoddy record, Musk had started to talk about "full reusability" for Falcon 9 rockets, about staging dozens of launches a year, and about flying NASA astronauts into space. Many thought the entrepreneur, still in his thirties, to be just another reckless self-promoter with a large ego. Privately, and sometimes publicly, they mocked the would-be disruptor. Musk's bold talk won critics not just among his competitors in the U.S. aerospace industry, but in important corners of Washington, D.C., space policy. Charles Bolden, who would serve as NASA Administrator during President Obama's tenure in the White House, was a self-described extreme skeptic of Musk and SpaceX. And the powerful U.S. senator who held NASA's purse strings, Richard Shelby of Alabama, declared that efforts to rely on private companies like SpaceX represented a "death march" for NASA.

"We cannot continue to coddle the dreams of rocket hobbyists and so-called 'commercial' providers who claim the future of U.S. human spaceflight can be achieved faster and cheaper," Shelby said in 2010.

Such skepticism fueled Musk, and indeed many of his employees. By Musk's reckoning, the launch industry snickering behind his back was ripe for disruption. Launching the Falcon 9 rocket would offer a suitable retort. Only a handful of nations had the capability to lift large payloads into orbit in 2008, and, with its new rocket, SpaceX aspired to join this elite club.

Whereas the Falcon 1 could lift a maximum of about 1,000 pounds into orbit, the Falcon 9 would be capable of pushing 23,000 pounds into space, putting it in line with the major rockets of the day. But even this was not enough to satisfy Musk. He demanded that each Falcon 9 be capable of flying multiple missions. He did not want to burn those nine brilliant engines for a few minutes, and then watch them fall into the sea, the standard practice for rocketry.

Imagine, Musk would often say, the first flight of a $100 million passenger aircraft. Suppose that at its destination all of the people onboard had to parachute out of the plane, with the jet subsequently crashing into the ocean. In such a world, air travel would be rare, dangerous, and prohibitively expensive. But that's largely how space travel worked.

Musk was far from the first visionary to understand that humanity's future among the stars could only be unlocked by launching rockets, landing them, and then rapidly reflying the same hardware for pennies on the dollar. Musk's genius is that he not only saw this future but believed in it enough to keep pushing, to keep fighting, and to will it into existence.

Now his team of talented engineers was taking the first steps toward that goal by firing up the Falcon 9 for the first time. Yet the sound whirling and tumbling and whipping over Mueller and a couple of dozen other SpaceX employees that night did not stop at the test site's fence-line, near the small Central Texas town of McGregor. Rather it rolled onward and outward, across the mostly treeless flatlands. This firing of the Falcon 9 rocket's first stage proved far louder than any previous test, and not just because its nine engines ran for nearly three minutes.

Days earlier a cold front had come barreling through, driving temperatures down near freezing. But by Saturday, November 22, 2008, a more temperate southerly flow returned, building a thick layer of clouds that swaddled the land like a newborn. When the rocket's nine engines ignited at 10:30 PM local time, the wind had died down, creating ideal conditions for the propagation of noise. Humid air absorbs less acoustical energy. The lack of wind allowed noise to spread undisturbed. And although sound waves do not bounce off clouds, their presence betrayed a temperature inversion, with warmer air above the surface. Because of this, as the Falcon 9 rumbled to life that night, its sound waves were bent back toward the Earth's surface, rather than continuing upward toward space. Such inversions are more common at night, which is why distant trains often sound louder after dark.

And so Central Texans saw and heard something that night that they never had before. The test rattled the windows and nerves of residents for twenty miles around. Some wondered if exploding bombs heralded the onset of World War III. Others felt the biblical end times had come and hid in closets. One child asked his mother if the Sun had exploded. Those who had actually heard of the SpaceX test site wanted to know just what the hell Elon Musk and his rocket company thought they were doing in the middle of the night. Even as Mueller and his propulsion team basked in the glow of success, panicked calls began to flood the local 911 operators.

In Southern California, the rest of the company's employees gathered around large screens showing the test, cheering as the Falcon 9 burned its engines. When calls of alarm started to flood in from Texas, however, SpaceX's senior leadership broke off their celebrations to respond to these concerns. "We certainly did not anticipate that reaction," said Gwynne Shotwell, the company's newly minted president and chief operating officer, who was also celebrating her forty-fifth birthday that night.

She had tried to warn local residents beforehand. Although the company had conducted more than 2,000 engine tests during the previous half decade, the vast majority fired up a single Merlin rocket engine for a much shorter period of time. Shotwell wanted Texans to know this one would be different, likely sparking more than the usual complaints of hot dogs rattling off grills or a cow's milk going bad. So SpaceX placed a notice in the local newspaper, the *McGregor Mirror*, and posted about the Saturday night test on the high school marquee. SpaceX officials had given interviews to local media. But all of this was to no avail as the community shook and an eerie orange glow spread across the horizon.

"We were up late that night," Shotwell said. "We were writing press releases to calm everyone down, explaining that it was not an alien invasion, that it was not the end of the world."

## Faster than anyone, too slow for Elon

The earliest years of SpaceX were truly desperate, as I recounted in *Liftoff* (William Morrow, 2021), my book about the origins of the company. Musk founded SpaceX to build a fleet of reusable rockets and spacecraft in order to, one day, settle Mars. This audacious goal had a simple first step, proving that a small team of engineers and technicians could build a basic rocket with a single engine and reach orbit at an affordable price. This meager task of building the Falcon 1 almost broke SpaceX. Down to its last chance in September 2008, the company enjoyed only the narrowest of successes on the fourth launch.

Afterward, SpaceX faced a difficult path. Though the Falcon 1 had finally reached orbit, few customers were lining up. The rocket was too small and had not yet proven itself to be reliable. The future, if SpaceX had one, lay in a bigger rocket and a larger pool of customers. The most important of these was NASA, which needed help shipping food and water to astronauts onboard the International Space Station. That fall, as SpaceX worked toward test-firing the Falcon 9 in Texas, NASA officials watched closely. And a mere month after the full-duration firing in late November, NASA awarded SpaceX $1.6 billion for a dozen cargo delivery missions. With this money SpaceX could complete development of the Falcon 9 and begin experimenting with reusing the first stage. But there was a catch: SpaceX would only receive the bulk of this funding once it started delivering cargo.

Money, therefore, remained tight. SpaceX raced to complete the Falcon 9 rocket, and the Dragon spacecraft to carry cargo, before the money ran out. Musk, the ringmaster, guided the company along this razor-thin line, and his methods were at times ruthless. Always, he urged his managers to spend less, and he pressured them to work harder and faster. But it was never fast enough. Originally he expected the launch-ready Falcon 9 to debut in 2008. In reality, his engineers and technicians worked miracles to achieve the first full-duration test-firing in late 2008. No one understood this better than Mueller, who spent countless nights and weekends away from his young family alongside his propulsion team. Even as the Falcon 9 brilliantly lit up that cloudy November night, Mueller understood it would never be enough.

"We were out there beating history, but Elon was still pissed at us," Mueller said. "Like everything else we've ever done, it was way slower than Elon wanted, and way faster than anyone had ever done it before. It was pretty much the story of our lives."

When Mueller dashed out of the control room bunker to see that first Falcon 9 firing, he left behind a key lieutenant hunkered over his computer monitors, looking at data. This trusted deputy was a young aerospace engineer named Kevin Miller. Whereas Tom Mueller felt pure joy as the rocket ignited, Miller simply felt relief. He bore responsibility for the nine Merlin engines and knew how much could go wrong as they ignited one by one. And if something did go wrong, Musk and Mueller would descend on him for answers.

During summers when he was growing up in Indiana, Miller would visit his grandparents in nearby Michigan. Every year, a highlight was a trip to the Michigan Space and Science Center. Though modest in scope, the museum boasted the Apollo 9 capsule that had flown into space in 1969 among its collection. The commander of the mission, Jim McDivitt, had gone to college in Jackson, and arranged for the small museum there to house the historic object. Miller loved peering into the capsule to see where McDivitt and the other astronauts had floated in space. Outside the museum, which has since closed, he would slowly walk around the monstrous F-1 rocket engine that stood on a concrete pad.

While he dreamed about space, Miller had no illusions about one day climbing on top of a rocket himself. A streak of Midwestern realism runs through him, and Miller recognized that since he got dizzy riding on a merry-go-round, microgravity was probably not for him. But as a boy gaping up at that F-1 engine and watching videos of it launching astronauts to the Moon, Miller became hooked. If he couldn't ride on big rockets, he would build them.

That goal had carried him to the in-state school Purdue, where Miller thrived and earned a coveted slot as a NASA "co-op" during his freshman year. This allowed him to spend part of his time working on real aerospace problems, with real engineers at the space agency. At the turn of the century, nearly all his classmates aspired to work at NASA, which had sent astronauts to the Moon and now proudly flew the space shuttle. After several co-ops, however, Miller realized a career in civil service was not for him. "I had met a lot of people who worked on things for twenty years and never saw anything fly," he said.

After earning a graduate degree in 2005, Miller looked toward the private sector. The dominant U.S. rocket propulsion company was Rocketdyne, but new entrants SpaceX and Jeff Bezos's Blue Origin offered an alternative pathway and more hands-on work. During his visit to Blue Origin in Washington, Miller faced eight hours of challenging technical questions in an academic-like setting. Although the engineering questions were difficult, the overall vibe was casual. SpaceX, then located in buildings scattered across El Segundo, California, proved far more chaotic.

Tim Buzza, the company's launch director, was too busy for a sit-down interview with Miller. So they walked and talked while Buzza oversaw work on a Falcon 1 stage being readied for shipment. Miller saw the rocket on the factory floor, with hardware scattered all about, as SpaceX worked toward the first Falcon 1 launch. Welders and machinists banged away in a machine shop. Seeing this beehive of activity drove home the difference between SpaceX and Blue Origin.

"It was pretty much night and day, what SpaceX was doing," he said. "Seeing the first stage on the floor sold me. They were really going to give this a go and build this rocket with just a few people."

Miller already had a job offer from Blue Origin. But his interviews with Buzza, Mueller, and others went well. He would be immediately helpful to the rocket startup and its small but growing propulsion team.

SpaceX offered him a position just before Miller's drop-dead date to let Blue Origin know. He accepted the SpaceX offer—a big part of the allure was a chance to work alongside Mueller.

Even then, years before SpaceX had any appreciable success, Mueller was highly regarded in the industry for his inventive thinking and technical adroitness. He had signed on to SpaceX as a founding employee with another engineer, Chris Thompson, back in 2002. Mueller oversaw propulsion, and Miller thought he would make a good teacher. After joining SpaceX in June 2005 as a development engineer, Miller worked alongside Jeremy Hollman, another great mentor who oversaw the Merlin rocket engine's development. Two years later, when Hollman left the company to start a family, Miller had earned the trust of Mueller and Musk. He was given responsibility for the most important piece of machinery at SpaceX, the Merlin engine.

## Falcon 1 to Falcon 5 to Falcon 9

Around the time Miller hired on, Musk began thinking seriously about a larger rocket, powered by more than the single Merlin engine that propelled the Falcon 1. Initially, he believed the logical next step for SpaceX was to evolve from a rocket with one Merlin to five, with one engine at the center and four clustered around it. But Musk changed his mind about that shortly after the debut flight of the Falcon 1. In the spring of 2006, the company started writing a detailed proposal to deliver cargo to the International Space Station for NASA. It became clear that the so-called Falcon 5 would simply not be powerful enough to meet NASA's requirements for multiple tons of water, food, and other supplies to be delivered by a single spacecraft.

Musk often makes important decisions in the middle of the night, hours after midnight. And, as far as he's concerned, the best time to communicate this decision is, of course, right then. So, one morning Musk called Chris Thompson in the wee hours. As the vice president of structures groggily took notes at his kitchen table, Thompson contemplated all the work they'd already done on the Falcon 5 project. It would all have to be scrapped. The following day he and Tom Mueller grumbled about the changes they would have to make to the thrust structure that supported the engines, to the propellant tanks, and to a hundred other major components.

"The next morning we were like, 'WTF is he thinking?'" Thompson said. "He was screwing all this crap up. But at the end of the day, he was the boss. Five engines, nine engines, whatever. Ten-foot diameter for the rocket, or twelve feet, doesn't matter to me."

At the time, Thompson and the rest of the SpaceX team were already drowning in the work of trying to reach orbit for the first time with the Falcon 1. Any plans beyond that might as well have been set on Jupiter. But Musk had the luxury of seeing the bigger picture, and he knew that a larger rocket would enable SpaceX to eventually launch humans into orbit. As he usually does when faced with a difficult choice, Musk went for the bolder option: nine engines instead of five.

It was a practical decision. Musk and the propulsion team had also considered building this second rocket with just a single, much more powerful engine. The basic idea was to scale up the Falcon 1's size and power. A single "Merlin 2" engine would be capable of putting at least ten times as much mass into orbit. But after some trade studies, this option was discarded. While Musk definitely wanted to develop a more powerful rocket engine, he understood this would be a major commitment. Building a more powerful Merlin would have cost hundreds of millions of dollars and years of development time. When SpaceX won the contract to deliver cargo to the International Space Station in August 2006, the space agency wanted deliveries to begin in 2010.

"We just didn't have the time or money to develop the Merlin 2 engine," Mueller said. "So Elon decided that instead of having one big engine, we're going to use a lot of Merlins. And we're like, 'Whoa, man. That's going to be tough.'"

Mueller well knew who would be putting in that tough work. He led the propulsion team by example, working late into the night alongside his engineers and technicians and approaching engine development with a curious mind. Rocket engine design involves a lot of tinkering to get the flow of propellants into a thrust chamber just right, to precisely control their combustion and channel the exhaust away. Mueller taught his team to pay attention to details and listen to the hardware for potential failures before they occurred. Mueller would engage with interns in the same attentive manner as he would with senior leaders at SpaceX, including Musk. His team loved him for this.

In the spring of 2007, the propulsion team had a big challenge before them. SpaceX's first Merlin engine could only be used for a single launch, as key components were charred when superheated gases moved through. With the Merlin 1C, SpaceX sought to upgrade the engine for multiple uses. They did this by embedding tiny channels within the thrust chamber and nozzle, using them to flow room-temperature kerosene throughout the engine as a coolant. This added complexity to the design, and the propulsion team struggled with clogged channels and cracked engine parts during tests. They kept burning through hardware as fast as it could be made.

But by mid-April, Kevin Miller and the other engineers were closing in on a full-duration test-firing of the new, regeneratively cooled Merlin 1C engine. Miller's life had settled into something of a routine by then, working part-time at the company's factory in California, and spending the remainder at the engine test site in McGregor, Texas. Every few weeks Miller and a handful of engineers and technicians would fly from Los Angeles to McGregor, running tests until the hardware gave out. When there were no more spare parts, Miller and the Merlin team would return to Los Angeles to regroup, fiddle with the engine, and plan their next test campaign.

They came close to completing a full duration test on a Saturday night, April 28, before their final nozzle failed. The next morning, ahead of his flight back to Los Angeles, Miller stopped by the blockhouse at McGregor to tidy up. Upon entering, Miller saw the test site director on the phone, his face turning ashen. After the call ended, Miller understood why. Musk had issued an unequivocal directive: they had to complete the full-duration engine test, without delay.

Such demands were part of the deal at SpaceX. Musk hired the smartest, hardest-working, and most creative people he could find. He provided fair compensation and generous stock options to give employees skin in the game. As they succeeded, the private valuation of SpaceX rose and their wealth increased. Some people joined because of this potential for financial reward. Others came because they truly believed in Musk's messianic mission to settle Mars and make humans a multiplanetary species. Working at SpaceX offered the best chance on Earth to make science fiction dreams a reality, and many employees relished being part of the journey. Still others knew that by putting in a few years at SpaceX they would learn a hell of a lot, earning a PhD from the rocketry equivalent of Harvard University. With SpaceX on a résumé, an engineer could write a ticket to anywhere in the industry.

Where Musk offered a lot, he also made wrenching demands. During crunch time on important projects, to work at SpaceX meant to live at SpaceX. An employee might go home for a few hours to sleep, but one's mind never strayed far from the job. Musk provided the vision, funding, and intense focus on building a reusable orbital rocket. He frequently led technical meetings and often made the most difficult decisions. And always, he pushed relentlessly forward, seeking to bring projects toward completion. But sometimes these project deadlines were arbitrary, driven by some external event rather than a reasonable schedule. This might be a public speech or presentation by Musk, and, when Musk wanted a flashy thing to share with the world, his team had to put in long hours to make it happen. For the Merlin 1C test, Musk had a meeting with space policymakers early the next week in Washington, D.C. Eager to show progress with his Falcon rockets, Musk intended to share video of the test.

Accordingly, Miller understood that Musk's demand was nonnegotiable. They would work until the job was done. Musk expedited the shipment of replacement hardware on his private jet to McGregor on Sunday. After its arrival, Miller and the engineers and technicians in Texas spent Sunday night preparing the reconfigured Merlin engine for the crucial test. After they worked straight through thirty-six hours, the Merlin 1C engine was ready.

It fired successfully for 170 seconds, and Musk got to give his triumphant speech the next day. SpaceX also issued a news release trumpeting the test. In his comments in this release, Musk even gave a nod to his exhausted team, saying "The success of Merlin is really due to the joint function of a great propulsion and test team."

Developing the new Merlin engine gave the propulsion team fits in other ways. For example, it would regularly spit out a part called a pintle injector into the open field at McGregor. A pintle design, which is similar in concept to a coaxial cable, feeds propellant into the engine's thrust chamber, with liquid oxygen running down the center and kerosene flowing around it in a separate tube. Calling this failure "punting the pintle," the engineers were confused because the design had worked on the original Merlin 1A engine. A few small modifications seemed to resolve the issue, but the root cause remained a mystery.

As ever, Musk was pushing his engine team to deliver the Merlin faster, and there were still so many issues to work through. With the Band-Aid fix in place, it would have been easy to consider the pintle-punting problem a closed issue. But instead, Tom Mueller shut down engine testing and worked with his engineers to fully understand the problem. He knew from long experience, shared with his team through stories about his hobbyist days and early SpaceX years, that unresolved small technical issues can become big problems on launch day.

So the team stood down for weeks, and Mueller took his lumps from Musk. A handful of engineers on the propulsion team were so flummoxed by the problems facing Merlin that they quit and left for Blue Origin. But all the while, Mueller's remaining engineers worked on these problems, including the pintle injector, and trusted their leader. Eventually they solved what could have been a fatal issue with the injector by designing a complex joint, with four different metal alloys squeezed into less than an inch between the copper pintle and aluminum injector.

"Even in the lowest times of engine development, my feeling was always that there was no technical obstacle we'd run into that Tom couldn't navigate us through," Miller said.

## Dog not scared

After SpaceX completed the full-duration Merlin 1C test in April 2007, much of the propulsion team's focus shifted to attaching one and then several Merlins onto the Falcon 9 rocket for additional testing. This was hot and dirty work, often performed high above the Texas scrublands on the facility's imposing, 135-foot-tall tripod. This concrete structure towered strikingly above an otherwise unremarkable stretch of countryside.

The nearby small town of McGregor is somehow located both in the middle of the state of Texas as well as in the middle of nowhere.

The tripod takes its name from the three massive legs on which it rises above the surrounding landscape. It stands 135 feet tall from its base to a landing platform, where engineers mounted the Falcon 9 rocket's first stage for tests. Visible for miles around, the tripod was a remnant of Beal Aerospace, a company founded in the late 1990s by Dallas banker Andy Beal, who had aspirations of becoming a rocket magnate. After Beal Aerospace flamed out in 2000, SpaceX acquired hundreds of acres at the McGregor property for tests of the Falcon 1.

During those early years the tripod went largely unused, not needed for the smaller booster. But it was sized properly for a bigger rocket like the Falcon 9. Because it stood so high, rocket engines could burn without the need to divert flame or exhaust, or concerns about damage to nearby buildings.

"The tripod is pretty kick-ass," Musk told me. "It's this giant booster stand, and I've been on that tripod a million times. It's a beast. There's a shitload of concrete and steel, thanks to Andy Beal. It was practically the only useful thing on the site when we got it."

In the early years, Musk made sure his employees knew the tripod would be used in the future. A classmate of Miller's at Purdue, Josh Jung, graduated in December 2003 and came to McGregor. Because he grew up in nearby Fredericksburg, Texas, Jung chose to call the test site home, and in January 2004 became the first full-time engineer to work there. Jung loved smoke and fire, and he brought this passion to SpaceX.

On his very first day on the job, Tim Buzza gave Jung the site tour. As they got to the tripod, Buzza gestured up to the tower and said that one day they were going to test big rockets up there. Jung just nodded along. At the time SpaceX didn't have a functional rocket engine, and cows were grazing in the grass nearby, milling about as if they owned the place. But sure enough, four years later, Jung was scrambling way up there with other engineers and technicians preparing the tripod for test-firings.

The tripod proved an intimidating place to work. Cold fronts during the fall, winter, and spring months would drop out of the north, howling across the exposed platform high in the sky. The tripod's platform, suspended more than 100 feet in the air, housed large doors like those in the bomb bay of a military aircraft. Only instead of opening to drop bombs, these sheet-metal plates would roll back to create an opening for the rocket's engines to expel their fire, heat, and exhaust downward, into the open air. When engineers and technicians needed to work on the rocket, such as while draining a turbopump after a test, these sheet-metal doors were of course closed. But this offered only minimal comfort.

"As you were crawling around under the engines on these bomb bay doors, they would move and ripple and dent," said Roger Carlson, a physicist who worked on Falcon 9 stage testing. "You would start to wonder whether they were going to hold your weight."

A technician who worked at McGregor, Cory Stewart, had taken in a stray, part black Labrador retriever that had shown up on-site one day, and named her Rockette. She faithfully followed Stewart everywhere, including up on the tripod. Atop this platform, with the bomb bay doors closed, Rockette thought nothing of scampering across the thin metal sheets. When someone asked, "Dog not scared?" the phrase caught on immediately. Soon, "Dog not scared" became the standard response when someone showed a bit of trepidation on the tripod. It became the SpaceX equivalent of "Hold my beer," used widely at the company as a way of saying yes to any seemingly impossible job or schedule challenge.

Even reaching the tripod's platform required a measure of fortitude. One option was to scale a ladder with more than 100 rungs in a steel frame cage. Or a worker could ride a construction elevator that climbed one of the three tripod legs. No one liked the elevator, though. It wobbled side to side as it slowly went up and down its single track, emitting a discordant shriek that was equal parts the tumbling of a large, unbalanced washing machine and a howling wild animal. The experience was all the more nerve-wracking because every time the Falcon 9 fired its engines, it violently shook the elevator and its components. During these tests the elevator was kept at the top of the tripod, because if left on the ground the engine exhaust plume would have badly burned it. After every test, the elevator would be lowered, lifted, and lowered again as a safety check before anyone would be permitted to ride it.

SpaceX had never had an on-the-job fatality, but that record almost came to an end because of the tripod elevator. Tom Mueller had responsibility for the site one day in 2008, when some welders were working on the tripod platform in preparation for the full-stage test-firing. Another welder was working in the bucket of a JLG crane that had been lifted up to the tripod's edge. Due to this ongoing work, there was a lockout-and-tagout safety procedure in place, which meant that the elevator was not to be used. However, Mueller said, the safety person on-site decided to take the elevator to the top of the tripod to check on the progress of the welders. As the elevator reached the top it bumped into the JLG crane, and the welder fell out.

"The only thing that saved his life is that he fell into that cage around the ladder," Mueller said. "He was lying there with a broken leg, all busted up, and unconscious. But if he hadn't hit that cage, he would have fallen all the way to the ground."

Risks aside, the view from atop the tripod could be sublime. The SpaceX facility sat within the boundary of an old, massive munitions facility built during World War II to help fuel the Allied efforts. The Bluebonnet Ordnance Plant, named after the Texas state flower, built four million bombs during its three-year life, some weighing as much as two thousand pounds. The government constructed more than 220 bunkers to store munitions during the war, and six decades later dozens still dotted the green fields and farms surrounding the tripod.

As part of preparation for stage testing, SpaceX hauled a 100-foot service structure to the top of the tripod to support the Falcon 9 rocket. The rickety structure included a stairway that topped out on a platform without any protective rails. Up there, the tall, thin structure would sway in the wind, leading to a dizzying, almost otherworldly effect. Employees referred to the steps leading to the very top of the tripod, more than 200 feet above the ground, as the "Stairway to Heaven."

It proved a good place to awe visitors and, sometimes, play practical jokes. When one of the company's new computer programmers, Robert Rose, came to McGregor, veteran launch engineer Ricky Lim offered to show him around. As they climbed the Stairway to Heaven, Lim explained that testing rockets in Texas afforded SpaceX much more freedom than in California. For example, Lim said, the company could easily test hypergolic fuels, which immediately catch fire when combined. In a cautionary tone, he added that employees should be prepared to evacuate immediately in case of a leak. How do you know there was a leak? Well, Lim helpfully explained, it has this characteristic rusty, burning, metallic smell. About a dozen steps later, Rose thought he smelled this very scent, and asked Lim if he did. In response to this question, Lim made no reply but to bolt down the stairs. Rose followed and found him two flights below, howling with laughter.

"Are we safe?" Rose asked, panting and nervous.

Lim stopped laughing long enough to explain that he was just messing with the newbie. The smell came from a pulp mill nearby. And then he started walking back up the steps.

For those fearful of heights, the Stairway to Heaven felt like the opposite of its name. Bulent Altan, an avionics engineer, thought he had conquered his fear of heights during the Falcon 1 years inside the basket of a lift. But on that exposed stairway, three times higher in the sky, his old fears came rushing back.

"I hated it," he said. "Think about a stairway that goes nowhere. It feels wobbly. It was windy and exposed. What really gets you is the rocket. You're standing there and this much larger and robust-looking thing is just moving in the wind, and that just freaks you out."

As senior director of avionics, Altan would sometimes visit McGregor to work with Catriona Chambers, an engineer who had responsibility for the first stage avionics and spent months on end in Texas. When something went wrong with the rocket's electronics, perhaps a power loss or controller issue, the avionics team would have to climb to the very top of the rocket, at the interstage, where the computer boxes were housed. During one of Altan's first visits to the tripod, Chambers, an old hand at the climb, led the way up the stairs. When she got to the small platform at the very top, she turned around to see Altan, crawling on hands and knees, at her feet. This big, burly, garrulous guy was just quivering on the metal stairs.

"What are you doing, Bulent?" Chambers asked. "Are you okay?"

He was not.

## "I barely breathe, or anything."

A single Merlin in 2008 produced 94,000 pounds of thrust by rapidly burning through oxidizer and propellant. And I do mean rapidly. If you were to drive a sedan nonstop from New York City to Miami, its internal combustion engine would consume about 350 pounds of gasoline over the course of a twenty-four-hour trip. Inside a Merlin engine, which could very nearly fit beneath the hood of most cars, a turbopump pressurizes and hurls 350 pounds of liquid oxygen and kerosene fuel into the combustion chamber *per second*.

Once ignited, this fuel produces a super-heated gas that is channeled through the thrust chamber and out of the nozzle, the cone-shaped exhaust channel that projects from the bottom of the engine. This gas blasts into the nozzle, traveling several thousand miles per hour. To channel the flow of this super-heated, supersonic gas, the nozzle must have strong walls, capable of absorbing an incredible amount of energy. This comes to about 10 megawatts, the average amount of power consumed by thousands of homes in the United States at any given time. And this nozzle is only about three feet across at its widest point.

"The Merlin rocket engine is a violent beast," Tom Mueller said.

Per Musk's directive, Mueller's propulsion team had to cluster together *nine* of these violent beasts as tightly together as possible, to fit within the twelve-foot diameter of the Falcon 9 rocket's base. Thousands of pounds of fuel, flowing into nine engines, producing superheated gas, all inside a tightly packed space. What could possibly go wrong?

The biggest problem, initially, was lighting the engines. This may sound fairly simple, but in reality, introducing just enough propellant, but not too much, at just the right time, proved tremendously difficult. There is a fine line between a controlled explosion—which is, basically, what makes a rocket fly—and a disastrous, uncontrolled explosion. When SpaceX first began test-firing the Merlin 1C engine at McGregor, engineers successfully ignited it at the right moment about 25 percent of the time. Sometimes after an abort, rocket fuel would leak into the engine and damage the nozzle. It was a dirty, messy, frustrating business.

And that was just one engine. "It was always like, man, how are we going to start nine of these things when we have trouble starting one?" Kevin Miller said. "We had gone through so many hard knocks with the Falcon 1, and that was a much simpler system. At times, it seemed like this was going to be insurmountable."

They started with one engine on the tripod in Texas. By early 2008, they had installed two, then three, then five, until finally there were nine engines on the stand. That summer, as the Falcon 1 engineers scrambled to save SpaceX with their fourth attempt to reach orbit from Kwajalein Atoll in the Pacific Ocean, the McGregor engineers were making progress on the much larger rocket.

The team confronted a seemingly unending sequence of challenges. An engineer's life consists of staring down one technical problem after another, fixing this or modifying that, in an interminable quest to make something work. Sometimes failure necessitates a wholesale redesign of a system. Then the whole pattern repeats itself: Design. Build. Test. Fix. Retest. Redesign. Retest. And so on. All in the hope that, one day, you will reach a point where everything, finally, more or less works. The right mindset for an engineer is to approach each problem as a puzzle, finding joy in the discovery of a solution. God bless them. For some of us non-engineers, a life of unending problems sounds pretty terrifying. Rocket science is *not* for us.

As the engineers and technicians in Texas progressed toward a vehicle with nine engines, one worry started to press upon Miller more than any other, above even his concerns about lighting the Merlins. He lost sleep over a part called the TVC actuator, a relatively small component at the top of the engine assembly that guides the direction of its thrust.

As you may remember from school, Newton told us that for every action there is an equal and opposite reaction. A rocket moves in the opposite direction of the flow of superheated gas expelled by its engines. The primary control for a rocket's direction, therefore, is the orientation of its engine nozzles. At launch, they are pretty much vertical. But during ascent, these nozzles will start to gimbal, or tilt, to change the direction of the rocket's thrust.

For each Merlin engine, two mechanical components physically moved the engine nozzle in the pitch and yaw dimensions (that is, side to side and up and down). These were known as thrust vector control (TVC) actuators. From the beginning of their development, Miller had worked with a supplier, Jansen's Aircraft Systems Controls, on the design of these actuators. As the Falcon 9 program advanced, the propulsion team kept having difficulties with them. The failure of one of the eighteen actuators in flight, leading to an offset engine, would not be catastrophic, as the other eight engines could compensate for the errant thrust. The problem is that when one of the actuators failed, it often would shove the affected nozzle "hard over," pushing it beyond the normal extension. This could be caused by something as simple as sensor error, or even a very tiny piece of debris as small as one-thousandth of an inch. In flight the nine Merlins were ballet dancers, moving in one direction, pirouetting, and then gliding back to their original position. But if one dancer fell, this would trip up the next dancer, and so on.

"The fear on the first stage was the close proximity of the nine engines and the nozzles," Miller said. "The clearances were tight. I was always afraid that we were going to ram an engine into its neighbor, or a wall. I was responsible for the engine, and I was responsible for the actuator. And if either one of those things went wrong and it caused damage to an engine, or we lost the stage, it was going to be a really bad day for me."

There were a million problems like this to work through. Throughout the summer and fall of 2008, Musk kept pressing the company to go faster. Everyone pitched in, from structures and avionics to software and test teams, as Musk was desperate for the full-duration test-firing to occur before NASA reached a decision on contracts for delivering cargo to the International Space Station. Seeing the Falcon 9 come to life for the first time would give NASA confidence that SpaceX was up to the task. But the propulsion team needed more time, as the engines still were shutting down prematurely.

As the Merlin delays mounted, Buzza started to worry about receiving a "silver bullet" call from Musk for the nine-engine test. This phenomenon dated back to the Falcon 1 days, on Kwajalein in the Pacific

Ocean, the remote atoll that served as SpaceX's first launch site. Though they worked with just a single Merlin engine then, there were still teething pains, with aborts due to fuel injectors, igniters, or other issues. Sometimes these aborts were spurious, but often they indicated serious problems, such as a gunked-up injector or another issue that, left unchecked, would be disastrous.

Typically the U.S. Army, which operated the range on Kwajalein, gave SpaceX two or three days to conduct a Falcon 1 test-firing or a launch. If the company missed that short window, they might have to wait another month to try again. The lost weeks frustrated Musk to no end.

"Elon thought we were being too conservative with our aborts, and there were times when he would want to go all in," Buzza said. "He told us he would take the responsibility to remove that conservatism. He said, 'When I tell you to let it go, turn off all the aborts and let it go.'"

Buzza and another launch engineer, Anne Chinnery, named this a "silver bullet." Chinnery went so far as to write test and launch procedures should Musk call in a silver bullet. In this scenario, the onboard flight computer was instructed to ignore all aborts coming from sensors on the rocket, be it a slightly high pressure or temperature or some other errant reading. It was the ultimate let-'er-rip move. "You shoot a vampire with a silver bullet," Buzza said, "You've got one shot."

On the night before the planned Falcon 9 test-firing, SpaceX engineers in McGregor were still refining their procedures. Because they were not running a silver bullet scenario, out-of-bounds readings by sensors would still trigger an abort. To ensure the test software correctly handled those scenarios, engineers ran endless dry runs of the countdown. At about 2 AM on the morning of November 22, said Josh Jung, they ran a scenario in which the software should have commanded the opening of water valves on the tripod to douse a fire. But instead, a single errant line in the software commanded every valve on the rocket, from propellant to igniter fluid, to open. Had this been a real countdown, in the real world, it might have destroyed the rocket. They fixed the bug and ran more tests.

"We ended up not going home that night," Jung said. "We just rolled up our jackets into pillows and slept beneath our desks for a few hours. It wasn't the first time we had done that."

The next morning, at about 7 AM, preparations began for the full-duration test-firing. Musk wanted the team to test early in the day, but of course that did not happen. Technical problems pushed the test from about ten o'clock in the morning until after ten at night. Through it all, Kevin Miller monitored the health of the nine Merlins. So much could go wrong. He tracked all of these possibilities in the data on his computer screens and in spreadsheets with detailed procedures. During critical tests he would almost invariably be pushed to the edge of consciousness in the final moments, consumed by the myriad possibilities for error and mesmerized by the flow of data about the rocket's health. "When we get down near T-0, I more or less black out," he said. "It's because I'm so locked in, looking for anything to go wrong. I barely breathe, or anything."

But that night the engines came on. All nine of them. Then the flow of propellant through the engines increased, with thousands of pounds gushing through the Falcon 9 per second. The fury of its output violently shook the Texas countryside and rattled the bunker. Miller could start to breathe.

He was just twenty-nine years old and had worked at SpaceX for only a little more than three years. Yet he already more or less ran the Merlin development program. And the engines were out there, in the humid Texas evening, kicking ass. "I thought about my co-op experience at NASA that night, where people worked on things that got canceled, and didn't get to see anything come to fruition," he said. "This was already my second rocket. It felt like a career-making achievement."

This was exactly what Miller had set out to do with his life. He had spent summers as a kid gazing up at history-making rocket engines in Michigan. Now, he was making some history of his own.

And yet there was still so much more to be done. SpaceX had a big rocket. It had come to life for the first time. But it still had to fly.

| 2 |

# LEARNING TO BE SCRAPPY

*January 10, 2009*
Cape Canaveral, Florida

Six weeks after its dramatic test-firing in Texas, a shiny white Falcon 9 rocket rolled out to its launch pad in Cape Canaveral, Florida. This seemed nothing short of a miracle, coming so soon after the first stage's hot fire test. But now here the rocket was, fully assembled and apparently ready to fly.

As SpaceX employees looked on at the launch pad, powerful hydraulic pistons slowly swung the rocket from a horizontal to vertical position, pointy end up. And there it stood, basking in the Florida sunshine, a beastly beauty. This milestone afforded those who had worked so hard to reach it a few minutes of relaxation and celebration.

Jubilant SpaceXers climbed one of the four lightning towers flanking the pad to gaze down on their creation and capture the moment in pictures. "It is pretty damn wild and scary up on top of those candlesticks, but it was one of the greatest days of my life," said Roger Carlson, who texted Musk a picture from atop the tower.

Musk, too, was eager to share his rocket with the world. He ordered large floodlights delivered to the site, and the company's communications director, Roger Gilbertson, spent hours arranging them for dramatic shots after dark. One of the resulting nighttime images captured the beams forming a large X across the rocket. As Carlson left the Air Force station that night, a guard suggested that perhaps the company should fly one of their rockets before putting on such a show.

But Musk was all about the show.

The miracle of that day is that it came so soon after SpaceX performed the full duration test-firing in Texas. After this test, Musk felt confident enough about the Falcon 9's prospects heading into the new year to tell reporters the rocket would be on the pad in Florida "in the next few months," preparing for a launch in 2009. Since then, the company had shipped the first stage from Texas, added a second stage, and topped the rocket with a bulbous payload fairing. Now, with a completed rocket on the pad in early January, it appeared that, for once, SpaceX had managed to keep to one of Musk's wildly optimistic schedules.

Musk's propensity for overambitious launch dates would later become known publicly as "Musk time," but his schedules were already legendary within the company. To ease the pain of Musk's demanding timelines, his team began referring to this as a "green lights to Malibu" philosophy. The drive from SpaceX's headquarters in Hawthorne to Malibu was about thirty miles, mostly along the Pacific Coast Highway. There were about a dozen traffic lights along the way, and if a driver were to hit every green light, and was willing to go about fifteen miles over the speed limit, the trip could be made in thirty minutes. With traffic, and hitting a majority of red lights, the drive could easily take an hour or longer. Essentially, then, it was impossible to reach Malibu in thirty minutes. Musk's schedules for SpaceX projects, however, invariably assumed green lights all the way to Malibu.

During the final weeks of 2008, the SpaceX team had scrambled to ship the first and second stages to Florida, and then spent the Christmas holiday assembling them on a concrete pad where a hangar would one day be built. As they worked long days and nights to integrate the first Falcon 9 rocket, Carlson came up with an impertinent nickname for the project: "Capricorn One."

This was a reference to a thriller movie that only the older hands at SpaceX would have seen in a theater, as it came out in 1978. Starring Elliot Gould as a reporter, and James Brolin, Sam Waterston, and O. J. Simpson of all people as astronauts, the plot centered around NASA's first mission to Mars, Capricorn One. Only the mission turns out to be fake, with the astronauts removed from the spacecraft just before launch. The movie's tagline was "Capricorn One: The mission that never got off the ground."

This was hardly the message SpaceX wanted to send about the Falcon 9, but the Capricorn One name nevertheless reflected the sentiment some employees felt about selling this rocket as a finished product. Because smashing though it looked on the launch pad, the Falcon 9 was not all it pretended to be. The first stage and its nine engines were real enough, but the rocket had almost none of the flight computers and other avionics needed for flight. The second stage was mostly hollow.

And the composite payload fairing on top, which protects a satellite on the ride to space, was also a mock-up that would never fly. The original Falcon 9 rocket, version 1.0, was designed to launch the Dragon spacecraft to the International Space Station. Dragon had almost exactly the same diameter as the twelve-foot-wide rocket it sat on. The composite fairing in Florida, by contrast, was more than sixteen feet across, to accommodate larger satellites. But the structure of version 1.0 of the rocket could not withstand the downward pressure exerted by that large fairing during launch. As the booster blasted upward, through the lower atmosphere, the loads on the fairing would crush the rocket. SpaceX had not even finalized the design of a usable payload fairing for the Falcon 9, so about half of the rocket displayed in Florida was fake.

No matter. After SpaceX raised the Falcon 9 rocket into a vertical position, reporters were invited to a news conference in the company's launch control center in Florida. With the rocket standing impressively tall nearby, the launch team spoke confidently about how SpaceX was performing important tests. The ground systems supporting and fueling the rocket prior to takeoff had checked out, Musk said. The transporter-erector that moved the rocket from its processing location to the launch pad, and then lifted it into a vertical position, worked perfectly. Brian Mosdell, SpaceX's launch site director; Tim Buzza, the vice president of launch operations; and Chris Thompson, vice president of structures, then took some questions.

At one point a reporter asked about the company's next steps to get ready for a launch. Buzza took the question and gave an honest answer. "It has some more paces to go through," Buzza acknowledged. After some additional testing in Florida, he explained that the rocket would be broken into pieces, which would be sent back to Texas or California for further work and tests.

This revealing comment had taken some of the shine off Musk's event, which was meant to establish how close SpaceX was to launch, not describe how the rocket would be removed from the launch pad and disassembled for shipping. "Elon was not happy about that, so he and I agreed that was my last press conference," Buzza said.

Musk was also not particularly happy when he found out about the Capricorn One sobriquet. He had taken the event really seriously. In his mind, shipping the Falcon 9 to Florida and standing it up on the launch pad represented meaningful progress. It showed the space community, and especially policymakers in Washington, D.C., that this scrappy little company from California could operate at America's oldest, and most storied, launch facility along the Florida coast. And that they had done meaningful work. SpaceX demonstrated that the Falcon 9 rocket fit with the infrastructure at Cape Canaveral, and they'd wrung out the hydraulics used to lift the booster. Given that, it's probably a good thing Musk never saw his team's mock movie poster, with images of Buzza, Mosdell, and Thompson superimposed over Brolin, Waterston, and Simpson.

"Once Elon heard about it, we had to stop using that name," Carlson said. "I guess he felt that Capricorn One was a little too flippant. And it was much more than a sales job. As a launch operations guy, I see the value of doing a pathfinder."

Carlson had meant no disrespect. Yet he, Buzza, and the others working in Florida also knew the truth. The Falcon 9 rocket standing in the Florida sunshine that weekend, bathed in lights at night, was not close to being ready to launch. Musk said during the news conference that the Falcon 9 rocket would launch that summer.

He would be off by a year, as SpaceX ran into some red lights on the way to Malibu.

## The future of SpaceX is blowin' in the wind

Amid the mad scramble to deliver the Falcon 9 rocket to Florida that winter, one of the first big jobs was getting it down off the tripod in Texas. This task, improbably, fell partly on the shoulders of Carlson, who had only recently joined the company. A physicist by training, Carlson, forty-one at the time, had spent the previous six years working at Northrop Grumman helping to test and integrate the James Webb Space Telescope. But by 2008 he could see that the oft-delayed project remained a long way from going to space, and he could not envision hanging around to see a payoff. (It would not, in fact, launch for thirteen more years.) Carlson applied to SpaceX that summer and joined the company days before the full-duration firing.

"I was basically hired to be the integration guy for launch, to make sure the rocket got fully assembled in the factory and shipped out to the launch site," Carlson said. "So I thought I was going to work in a factory, never see a launch site, and never travel."

During his first week on the job in California, Tim Buzza threw Carlson a curveball. The company needed to get the Falcon 9 rocket's first stage down off the tripod in Texas, so it could be shipped to Florida. Buzza recalled that Carlson had worked with cranes before, with the Webb telescope, and asked him to travel to McGregor to help. Carlson did have experience with cranes, but those had been indoor bridge cranes, a specialized type used in controlled environments. Such cranes are connected to the building and are used to lift heavy objects in clean rooms. Carlson had never worked with an outdoor crane. But he was the new guy, ready to help any way he could.

It proved to be no simple task. At the base of the rocket was a thrust structure, a square metal frame that housed the Merlin engines. It looked something like a tic-tac-toe board, with each box supporting an engine. On top of this thrust structure was a "run" tank, which held the kerosene fuel and liquid oxygen propellant. This run tank had thicker and more durable walls than the flight version of the tank. The flight tank could be more delicate, with less mass, as it was not designed to take dozens of refuelings and engine firings.

After the test-firing, to get the Falcon 9 first stage down from the tripod tower, SpaceX planned to use a large crane to lift the run tank off the top of the thrust structure and set it on the ground. The crane would then pick up the flight tank and raise it to the top of the tripod, where technicians would secure it to the thrust structure. After the two sections were bolted together, the first stage would be lifted down from the tripod onto a waiting semitrailer bound for Florida.

There was one big flaw in this plan: the wind. At the top of the tripod, conditions were often gusty, and the wind blew even harder another 100 feet up at the top of the stage. It was nerve-wracking for engineers and technicians to work up there, hopping from the top of the Stairway to Heaven onto the rounded top of the tank and back.

"I wouldn't say I was afraid of heights before coming to SpaceX, but I was certainly cautious about heights," Carlson said. "By the time I was done at SpaceX, I could have been one of those guys walking around on the steel girders, building the Empire State Building."

SpaceX hired Wales Crane & Rigging Service for the task. The company brought what it advertised as the second-largest crane in Texas to perform the delicate operations in the sky. Initially, SpaceX tried to wait out the wind. Carlson and the other engineers held up operations for a few days, trying to find optimal conditions in which to work. Finally, they decided to try in the middle of the night when winds typically were lowest. Late one night in November, the skilled crane operator and technicians did manage to remove the run tank and bring it down to the ground. It had been a close call, however, as the large tank swayed in the wind like a billowing sail on the seas.

During this operation Carlson and the others realized they were not going to be able to lift the Falcon 9 flight tank and attach it to the thrust structure. That process was a bit like assembling a piece of IKEA furniture, with pins and holes. Only in this case there were 144 steel rods, about as thick as an index finger, that connected the engine section to the flight tank. Each of these rods had a tight clearance around the thrust structure and its nine engines, and had to line up just right. With the flight tank dangling from a crane and blowing in the wind, it soon became clear that this kind of precise, delicate work was impossible.

So they improvised a backup plan. This entailed bringing down the thrust structure with the engines and shipping it to Florida separately from the flight tank. This would not be overly difficult, as the thrust structure was dense and less prone to swaying in the wind. However, it created a new problem. Rocket engines are complex machines, built with the least possible mass. Therefore, while their performance is robust, the engines themselves are not designed to carry weight. So the thrust structure could not be set on the ground because it might damage the engine nozzles that protruded from its bottom. The structure needed a cradle, rather like a carton of eggs, its framework carrying the weight at the thrust structure's hard points. The problem was, SpaceX had no such cradle on which to set the thrust structure, because no one had anticipated taking the engine section down separately. So as darkness closed in, the thrust structure hung off the end of the second largest crane in Texas.

"We had been working all night to lift the run tank down and worked all day to bring the thrust structure down, and now it was about eight o'clock at night," Carlson said. "We have all nine engines hanging from a crane with no way to put them down. I was the lead for this operation, and I remember standing there just feeling really alone. All of SpaceX was hanging in there from a crane in front of me, the most valuable nugget that the company had."

Carlson had been at SpaceX for only a few weeks. He'd been hired, he thought, to assemble rockets at the SpaceX factory in Hawthorne. But he found himself 1,400 miles away, in the middle of nowhere Texas, stressed beyond his wits. So it goes at SpaceX. You're hired. You do whatever job is put before you. And often it involves great responsibility. So Carlson and the other engineers began making phone calls. They arranged to have some welders arrive on-site the first thing in the morning to build a custom cradle. He and others would then stand vigil around the crane overnight, just to ensure that the brake holding the engine section in place did not fail, or that the crane's hydraulic fuel did not leak.

It was a long night.

The experience taught Carlson a lot about SpaceX and how fast the company moved. At Northrop, building a custom cradle would have become its own mini-program with design reviews, taking months to build rather than hours.

## Conflict at the Cape

The acquisition of that Florida launch pad had not come easily. At the time SpaceX began developing the Falcon 9, it possessed only a single launch site, on a tiny islet in the central Pacific Ocean. This would not do for a larger rocket, nor would NASA contract SpaceX to launch cargo missions from the middle of nowhere.

To reach the International Space Station, the Falcon 9 needed to launch from a spaceport facing eastward, and while there were a handful of options in the United States, Musk wanted a site in Florida at Cape Canaveral. For a company that aspired to one day send humans to space, this was really the only choice. The Cape was also near key facilities for NASA and the U.S. Air Force, which would both be critical customers for the company's medium-lift rocket. But could SpaceX crack the exclusive club at the most hallowed spaceport in the world?

In the immediate aftermath of World War II, the U.S. military had chosen a 16,000-acre barrier island on the east coast of Florida as a test site for missiles. The location was ideal—about as far south as one could get in the Lower 48 states in a relatively unpopulated area; accessible by rail, shipping, and roads; possessed of mostly favorable weather; and facing out over an ocean with thousands of miles of water into which spent rocket stages could fall harmlessly. In the decades after the war, the Air Force had built dozens of launch pads there for various purposes, from missiles and suborbital rockets up to much larger rockets. In 1962 NASA set up shop on an adjacent spit of land, Merritt Island, to lay down the pads that would launch humans to the Moon.

Musk started negotiating with the Air Force to acquire one of those older pads in 2006, and his focus soon turned to a large site near the northern end of Cape Canaveral. Known as Space Launch Complex 40, or "Slick 40," the facility had launched Titan rockets from 1965 until 2005, when the aging fleet was retired. After Titans stopped flying, the Air Force no longer needed the pad. Musk coveted the site because the Titan IV had been a heavy-lift vehicle. Although the obsolete launch tower was demolished and sold for scrap, the Air Force left behind a sturdy concrete deck that could support Falcon 9 operations, a large "flame trench" to carry rocket exhaust away from the booster, and four large lighting towers. Constructing all that from the ground up would have cost SpaceX about $100 million, so the complex represented a huge savings for Musk, who wanted to spend a tiny fraction of that for the entire site.

But not everyone wanted to turn the historic launch pad over to the upstarts. Foremost among these opponents was Lockheed Martin, which had launched the Titan rockets from SLC-40 and did not relish abandoning the pad to Musk. After SpaceX applied for a license in 2006, an extensive lobbying campaign ensued to block the transfer. Although officials from Lockheed and Boeing did not expect SpaceX to ultimately succeed with the Falcon 9, they were hedging their bets. Additionally, they resented Musk for his ego, his money, and his talent. He was not deferential, but brash. He had these odd origins in South Africa. Essentially, they argued, do you really want to let *this guy* onto the holy grounds of America's largest and oldest spaceport?

Much of this pressure fell on Susan Helms, who commanded the 45th Space Wing. This Air Force unit oversaw all aspects of safety and launch operations at Cape Canaveral, and assisted NASA with Kennedy Space Center. Legally speaking, Helms had total authority to approve SpaceX's application to lease SLC-40. But the reality was more complex.

Helms attended the U.S. Air Force Academy and later became a NASA astronaut. She had a storied career at the space agency, including five shuttle flights, setting a record for the world's longest spacewalk, and becoming the first woman to live on the International Space Station in 2001. After retiring from the astronaut corps, Helms returned to the Air Force and worked on the space side of the military. She first met Musk in 2004, and she said he was different from other would-be space entrepreneurs.

"I think what set him apart from the others that we were speaking to at that time was the fact that he actually had an executable plan," Helms said. "He didn't have a marketing pitch. He had an executable plan of how he was going to build a rocket and get to operations."

Helms took command of the nation's top spaceport in June 2006. The primary tenants at the Cape then were Lockheed and Boeing, which were in the process of forming a jointly owned rocket company called United Launch Alliance. "It was a collegial environment," she said of the military's relationship with the traditional launch companies. "But when Lockheed and Boeing picked up on SpaceX's interest, the antibodies came out."

Helms understood that the decision of whether to lease SLC-40 to SpaceX would become a political one. The traditional powers wanted to box SpaceX out and were lobbying senior Air Force officials accordingly. While she recognized that SpaceX had some maturing to do, Helms's interactions with Musk convinced her that he was serious about reaching orbit. She was inclined to approve SpaceX's application, but with the mounting political pressure, Helms wanted cover from the brass above her. So she went up the chain to her four-star general, Kevin Chilton.

He was also a former NASA astronaut and had responsibility for space operations within the Air Force.

"There were phone calls made to try and stop my decision, above my level," she said. "General Chilton was on the receiving end of some of those calls. People were basically advising him not to have SpaceX get the SLC-40 lease. I had been talking with the Florida Spaceport Authority, and they were all in on Elon coming to Cape Canaveral. The only people who were really against it were his competitors, and I think you probably know who those are."

The negotiations and lobbying persisted into 2007, until a decision had to be made. The key moment came when Helms met with Chilton that spring. It was pure happenstance that two astronauts were in the chain of command for the lease decision, but this played to SpaceX's advantage. Helms explained that the only opposition to the lease came from companies that might one day have to compete with SpaceX. This did not seem fair, she reasoned. As astronauts, both Helms and Chilton had spent more than a decade at NASA. This gave them a broader view of the space industry than someone brought up solely through the Department of Defense. This vantage point helped them see how SpaceX might disrupt a stagnant launch industry.

"I told him that as far as the Wing was concerned, this was for the good of the Wing, and the good of the country," Helms said. "And General Chilton also believed that SpaceX was going to shake up the industry in a way that the government would find advantageous. He agreed to back me up."

After this meeting, Helms decided to sign the lease. Before doing so, she made two phone calls. The first was to Musk, who expressed appreciation. The second was to Mike Gass, the chief executive of United Launch Alliance.

The first thing Gass said in reply was, "I am very disappointed." He did not say much else after that.

Helms and other officials joined Musk for an official groundbreaking at SLC-40 on November 1, 2007. Much to the chagrin of the nation's dominant rocket company, SpaceX had wedged its foot into the door at the most prestigious launch facility in the world.

## SpaceX learns to be very, very scrappy

The launch site developed slowly for the first months. But in February 2008, SpaceX found the person they needed to accelerate progress: Brian Mosdell. He was an "old space" guy, having worked at the Cape since 1991, building pads and launching rockets for the leading companies of the U.S. defense industry, including Boeing, Lockheed Martin, and General Dynamics. And Mosdell had the local knowledge SpaceX desperately needed. In the early 1990s, he had helped upgrade SLC-40 for a version of the Titan rocket with a Centaur upper stage. Additionally, Mosdell had a healthy respect for safety procedures: in 1991, he had fallen into a fourteen-foot-deep well at the site and nearly died.

He had also launched rockets from Florida. When Boeing and Lockheed merged their rocket businesses in 2005, Mosdell became chief launch conductor for both the Atlas and Delta rockets. These were the workhorses of the U.S. military, launching spy satellites, GPS spacecraft, and important science missions for NASA. SpaceX, by comparison, was a nobody.

One of Mosdell's former engineers at Boeing, Neil Hicks, had gone to work at SpaceX in 2007. Throughout that year he kept calling, telling Mosdell SpaceX could use his talents. "I was flattered, didn't have any serious interest," Mosdell said. "I had, I think most people at that time had, a general misperception that SpaceX was a PowerPoint company." That is, it consisted of flashy marketing presentations and press releases, and little else.

In January 2008 SpaceX invited Mosdell to visit the company's new headquarters in Hawthorne. He thought it might be fun to visit California for a few days, and he'd started to have some concerns about working at United Launch Alliance. When he sat down with Musk, the conversation quickly turned to the Delta IV Heavy rocket. It was powered by a new, American-made rocket engine, the RS-68. Musk was curious about the near failure of the rocket during its debut three years earlier. It had narrowly averted catastrophe after its three main engines shut down early due to a sensor failure. The mishap investigation dragged on for years, and the Delta IV Heavy had only just made its second flight. The lengthy approval-by-committee process among Boeing officials and the Air Force had driven Mosdell nuts. These frustrations made him more attentive to Musk's offer.

Mosdell had a job at SpaceX if he agreed to wear two hats, that of chief engineer of the launch site in Florida as well as its director. It would be a lot of work, but Mosdell had been intrigued by SpaceX. During the visit to California he'd seen what he deemed to be at least $25 million in flight hardware, shredding his conception of SpaceX as existing only on paper. He also liked the people he met, and believed they knew their business. So he became the tenth employee in Florida. Musk's instructions to Mosdell were simple: build a launch site as quickly as possible, for the least possible amount of money.

Mosdell started by creating a master plan to move from that first day of work on the pad through an initial launch. Meticulously, Mosdell laid out the staffing requirements he thought he would need, and cost estimates that could be rolled into a budget. From Florida, he would join weekly meetings with SpaceX Vice President of Launch Tim Buzza and the company's other senior leaders back in Hawthorne. During one of these first meetings, Buzza invited Mosdell to speak about his plans.

The directors, seated around a speaker phone, listened as Mosdell discussed his master plan and explained how he planned to "lay out a budget." After this, Mosdell paused, and there was silence. Then he heard laughter on the other end of the phone. He wondered if he had said something stupid. When the laughter subsided, he asked what he had said to provoke it. The other directors explained that at SpaceX, they did not create budgets. They just went and executed.

That is not to say SpaceX was careless with costs. Quite the contrary, as Musk challenged his directors to be as cutthroat about expenses as possible. "Scrappy was a big word," Mosdell said. "Back then you had to be scrappy. So when you laid out your plans they had to be scrappy. If everyone in the approval cycle felt they were scrappy enough, you could go for it. If not, they threw you out of the room."

Such a philosophy lay worlds apart from the staid business environment Mosdell had operated in for nearly two decades, working for large defense contractors. When NASA or the Department of Defense wanted a new rocket, they would go to industry and solicit a design. Then, for whichever company won the contract, Uncle Sam would pay for all development costs, including overruns and any modifications to the launch site. Therefore, the companies Mosdell worked for did not blink at high prices. In the early 2000s, Boeing had modified a site in Florida for its new Delta IV rocket, spending $375 million. When the Air Force decided it must also have the capability to launch Delta IV rockets from California, it spent $500 million modernizing a pad at Vandenberg Air Force Base (now Vandenberg Space Force Base).

This largesse extended to other government agencies and spaceflight. In 2005 the U.S. defense agency responsible for building and operating spy satellites, the National Reconnaissance Office, built a new facility to process satellites for launch vehicles at Cape Canaveral. Ultimately, it built a sparkling new 810,000-square-foot facility with one bay as high as 200 feet. The Eastern Processing Facility ended up costing an estimated $2 billion.

As SpaceX set about to revamp its newly acquired launch site at Cape Canaveral, including the launch pad, ground systems, and a processing hangar, Mosdell was not given a budget of hundreds of millions of dollars. Rather, Musk told the launch site team to build everything for just $20 million.

So they were S-C-R-A-P-P-Y.

To begin with, the SpaceX processing facility—where the company would take delivery of a customer's payloads, test them, and attach them to the rocket—was not going to be as posh as the nearby $2 billion facility operated by the National Reconnaissance Office. Rather, SpaceX reached out to a prefabricated building company in Fort Lauderdale, advertised on local radio by Dan Marino, the famous former Miami Dolphins quarterback. The company, United Steel Building, assembled a simple, steel-framed structure with some basic siding in a couple of months at the Cape. Then SpaceX employees installed electrical systems, lighting, and a basic air conditioning system. The company's processing hangar would cost less than $1 million.

Sometimes Musk's demands for building-on-the-cheap were over the top, and directors had to find ways to manage around the boss. For example, the hangar's concrete foundation needed steel rebar for strength. Mosdell said Musk would not approve the purchase order due to cost. The launch site team found cheaper materials from China, but Musk still refused it, questioning the need for rebar at all. "Any civil or structural engineer knows that's ridiculous, so we just broke the purchase orders down into smaller amounts, below his approval level, and moved on," Mosdell said.

Another way Mosdell and the Florida team proved their scrappy mettle was through buying scrap metal. Rockets require tons—literally tons and tons—of gases and liquids stored at high pressure to fuel and support the loading of propellant onto the vehicle, and keep it primed for launch. These "consumables" are stored in pressure vessels, which have multilayered walls about six inches thick. For example, nitrogen is used frequently to purge rocket engines to prevent the accumulation of volatile mixtures of gas before ignition. Nitrogen and other consumables are stored in very heavy tanks about thirty feet long and six feet in diameter. New pressure vessels cost about $3 million, and several were needed for various Falcon 9 consumables. Buying just a few would have blown the entire launch site budget.

So SpaceX nosed around abandoned launch pads to find scrap pressure vessels from old rocket programs. The company sourced them from old launch sites at Cape Canaveral and NASA's nearby launch pads at Kennedy Space Center, as well as from the old Santa Susana Field Laboratory located in the Simi Hills west of Los Angeles, where the upper stage engines were tested for the Saturn V rocket decades earlier. A NASA official working with the company on the commercial cargo program, Kathy Lueders, played a crucial role in facilitating these deals.

Two SpaceX employees, Chuck Wagner and Kary Policht, prospected abandoned launch sites with such zeal that the pair earned the nickname "Sanford and Son" from Air Force officials they worked with, a reference to a 1970s television show about a cantankerous junk dealer. Among their most important reclamation projects was acquiring a massive spherical tank located at Launch Complex 37, about two miles down the road from SLC-40.

The site had been used decades earlier during the Apollo Moon program, with the tank storing nitrogen. Mosdell had led propulsion ground system design and construction activities there from 1997 to

2002, reconfiguring it for the Delta IV rocket. At the time he suggested repurposing the old nitrogen tank for Delta launches. "That idea died a quick death," Mosdell said. Boeing managers told him it would be too difficult to perform a structural analysis and certify its use for modern launches. Leave it alone, he was told.

The LOX ball is moved to SLC-40 after SpaceX bought it as scrap. | PHOTO CREDIT: TIM BUZZA

Standing about sixty-five feet tall, the tank was sized such that it would be ideal to store liquid oxygen, or LOX. Buying a new LOX ball capable of storing 100,000 gallons would cost about $3 million, and Musk had already refused a purchase order for a new tank. So Mosdell knew SpaceX would have a different attitude than his former employer. Helms was happy to transfer the tank, which was on the Air Force's "scrap list." As there was no value associated with it in the military computer system, SpaceX eyeballed it at $86,000. The Air Force estimator agreed, and SpaceX had to pay $1 over this value. So Buzza had the company wire $86,001 to the government overnight, completing the purchase on a Friday. The next morning a local rigging and hauling company, Beyel Brothers, loaded the massive tank onto a specialized transporter and drove it down the road to the SpaceX launch site. In the weeks after the tank was removed from Launch Complex 37, some Boeing commuters complained of missing the turn into their site because they had grown so accustomed to the giant sphere as a landmark.

SpaceX still had to certify the aged tank for present-day use. An external analysis identified multiple spots of rust and areas of concern for fatigue or cracking of the tank's structure. Someone would therefore need to go inside to verify the integrity of the tank and clean out any debris. No one really wanted to do this task. That is how one afternoon two of the company's youngest engineers, Tyler Grinnell and Chris Wallden, found themselves standing on a steel catwalk built atop the tank, about seventy feet in the air. Some of the handrails were rusted.

It was probably not the safest thing SpaceX ever did, Mosdell said. Nevertheless, the two interns were lowered into the tank from a winch, wearing harnesses. SpaceX had taken some precautions, such as placing spotlights, dropping an air monitor to ensure there were no dangerous fumes, and pumping in fresh air to ensure good circulation. As they descended, Grinnell and Wallden took pictures, performed visual inspections, and picked up debris. All told, the testing and analysis cost SpaceX about $200,000. For a tenth of the cost, in a tenth of a time, the company had its liquid oxygen storage in Florida. Hundreds of launches later, the LOX ball is still working just fine.

A few months later Musk visited the launch site to film a promotional hype video for the company. During the recording, which featured renderings of a Falcon 9 launch, Musk and Mosdell walked around the site, with the LOX ball featuring prominently. As the camera panned to show the pair standing on the catwalk, Musk gleefully exclaimed, "Here we are, on top of our giant ball of liquid oxygen."

Then he delivered the kind of punchline that was pure Musk. He is at times brilliant, and at times foolhardy. He can be exhaustively demanding and censorious. But sometimes, Elon Reeve Musk is just a goofy kid with a puerile sense of humor.

"They say SpaceX has big balls," he quipped. "And it's true."

## "Don't listen to her, she is crazy"

As they scrounged materials for their launch site, SpaceX officials also had to learn to work with the buttoned-down brass that operated Cape Canaveral. The Air Force's 45th Space Wing, led by Helms, bore responsibility for range safety, which meant ensuring that any company launching a rocket took precautions to protect people and assets on the ground. They took this job seriously and were averse to change.

The traditional government contractors who used the range were well accustomed to this bureaucracy and familiar with the military's rules and regulations. Because companies like United Launch Alliance were flying payloads for the U.S. government, and received reimbursement for all costs, they accepted and followed every one of these rules as a matter of course, regardless of expense.

"The government is a massive bureaucracy in this respect and had jillions of regulations and rules about how rockets had to be built, and the public safety criterion," Helms said.

Musk typically views regulations as unnecessary impediments, and at the very least, he told his directors to question all rules and requirements. For her part, Helms recognized that SpaceX operated in a different manner than other contractors. She pushed her team to find a way to work with SpaceX while also ensuring the safety of launches, meeting the company halfway. She urged her safety officials to find ways in which the company could meet the "intent" of the range regulations, but not have to reach the exacting standards called out line by line.

For example, prior to every launch, ULA meticulously prepared a 500-page document known in military jargon as a "Universal Documentation System." This guidebook laid out, in mind-numbing detail, every service the range would provide, such as communications or weather forecasting, and when they would do so for an individual launch. Essentially, the document provided marching orders to the range for its operations ahead of a launch, said Mike McAleenan, a forecaster with the 45th Weather Squadron assigned to SpaceX. However, his colleagues at SpaceX did not want to spend the time crafting such a document for every launch. Instead, the company preferred to create a generic document and modify it slightly for each mission.

"The range was not happy about that, but SpaceX pushed it through," McAleenan said. "And they're still using the original document, with very few updates. Meanwhile, ULA still does 500 pages for every single launch they do. It's amazing that it hasn't changed. But from the beginning SpaceX set themselves up to launch every four days. They were always pushing to simplify things."

In the early years, some range officials often pushed back on SpaceX and Helms's accommodating leadership. As Helms sought to find a path to "yes we can," not all of her range safety team felt the same way. During one of the first meetings between the range and SpaceX directors, an odd incident happened. Hans Koenigsmann, the company's vice president of avionics, was seated in the middle of the table next to a senior range safety officer. As Helms spoke of "breaking glass," this old guard official slipped a note on a folded piece of white paper, about twice the size of a Post-it note, to Koenigsmann.

As he read the note, Koenigsmann was incredulous. "Don't listen to her, she is crazy" was written on the paper. The "her" referred to Helms and evidenced the apparent contempt some officers felt for her leadership. As he looked up, Koenigsmann's first thought was, what if Helms called on him, and asked about the note? For a panicked moment, he actually contemplated eating the paper, to get rid of the evidence. But he realized that would be conspicuous. So he stuck the note in his pocket and threw it away after the meeting. "I felt horrified," he said. "It was just a very awkward position to be in."

Fortunately for Koenigsmann and the rest of the company, the note-passing official was soon replaced. An engineer named Howard Schindzielorz became chief engineer of the safety office, and he proved more amenable to the changes Helms sought to implement. By the time she left Florida for her next post in October 2008, the SpaceX team had reached a workable relationship with Schindzielorz and his safety team.

Other tenants of the Florida spaceport were similarly hostile, and less circumspect. During the Capricorn One era, SpaceX had not yet built a hangar. So the techs and engineers working on the rocket in early 2009 did so out in the open, exposed, on the hangar's foundation. ULA employees would drive up the access road and stop at the fence. They took photos, shouted mocking comments, and laughed at their would-be competitors. SpaceX eventually placed a screen around the chain-link fence to block the view from the road, and Helms stepped in with a cease-and-desist order.

Helms took other steps to ensure that Cape Canaveral accepted SpaceX into the community, willingly or not. During monthly meetings, the launch companies who were operating on the range would gather with officials from the Air Force and National Reconnaissance Office. Helms very pointedly saved a seat next to her for SpaceX VP Tim Buzza, sending a message to the big dogs on-site that SpaceX had earned its place in Florida and should be treated accordingly.

"She was instrumental in helping us get through the bullshit in the early years," Buzza said.

Since NASA had hired SpaceX to provide cargo delivery, the space agency was generally more welcoming. The senior official who oversaw human spaceflight, Bill Gerstenmaier, made this welcome clear by accepting that SpaceX did things differently. He was an important ally, as any NASA program that touched on human spaceflight ran through Gerstenmaier. He made the key decisions about major contracts at NASA and commanded a budget that approached $10 billion a year.

On Thanksgiving Day 2009, about a year after NASA awarded SpaceX a contract to deliver cargo to the International Space Station, Gerstenmaier came to Florida to check on SpaceX's progress. At the time, the space shuttle had just five remaining flights scheduled, and NASA would need the private companies it had hired to start delivering food, water, and science experiments to the International Space Station soon. Gerstenmaier wanted to see everything, so Mosdell took him all around the pad, into the sea van containers, and on top of the propellant tanks to explore all of the materials SpaceX had salvaged from around the Cape. "He asked a million questions," Mosdell said.

As they toured SLC-40, Gerstenmaier related some of his experiences in Russia. He had lived in the country in 1996, supporting the flight of NASA astronaut Shannon Lucid to the Mir space station. Later he had led the International Space Station program for NASA and was now negotiating with Russia for access to Soyuz seats, which NASA needed to buy for its astronauts once the shuttle stopped flying. As part of this work, Gerstenmaier had spent weeks at a time in Baikonur, Kazakhstan, the dusty, World War II–era cosmodrome launch site for Russia's human missions.

Near the end of the tour, Mosdell took Gerstenmaier up on top of the LOX ball, onto the steel catwalk. Mosdell felt sheepish as he pointed out some rust and explained that SpaceX had tested the structure to ensure its robustness. Gerstenmaier, who is soft-spoken, just smiled.

"This is nothing compared to what I'm used to out of Baikonur," Gerstenmaier explained. "I've had to walk the gantry, from the service structure over to the rocket, across a boardwalk of wooden planks. Some of which were missing."

Gerstenmaier had a wicked sense of humor. During a later visit to Cape Canaveral, he spied Buzza and Koenigsmann walking down a street south of the spaceport. It was a semi-regular moving day: SpaceX-ers would periodically transfer from one hotel to another, or to a condo, during their weeks-long stays in Florida working on the rocket and the launch site. They were forever looking for a slightly better room, perhaps closer to the launch site, or with a superior breakfast buffet. On this day Buzza and Koenigsmann had put their clothes onto a luggage cart and were rolling it down a busy Florida roadway. Shambling down the sidewalk with these ragged belongings, the pair might have been mistaken for homeless people.

After recognizing the two engineers, Gerstenmaier slowed the car and rolled down his window. "Must be tough times at SpaceX," he quipped, before laughing and driving off.

And in fact, setting up the Falcon 9 launch pad had been rough times for SpaceX. But launching would be more difficult still.

| 3 |

# FLIGHT ONE: REACHING ORBIT

*September 28, 2008*
Bend, Oregon

While working late one evening, Robert Rose took a break to scan news headlines on the internet. A programmer, Rose wrote code for the video game industry, including the *Syphon Filter* series of third-person shooters for Sony's PlayStation. As he waited for his code to compile, Rose saw that a company named SpaceX had just successfully launched its first rocket. He clicked through and found a video of the launch. It was the first time he had seen continuous video of a rocket from liftoff all the way to orbit.

Then he watched the video again. And again. "I thought it was the coolest thing ever," he said.

Rose then browsed over to SpaceX's website. When he checked the job listings, the very first one sought a flight software engineer with skills such as the C++ programming language. He was qualified. He liked space. And so he thought, *Why not me*? Rose sent in a cover letter and a résumé, and a few weeks later he found himself in Hawthorne for a job interview.

At the company's headquarters, Rose spoke with several SpaceX employees who asked him very difficult technical questions. Because Rose impressed during these interviews, his recruiter coached him for the last step of the process—meeting with Musk. This was to ensure Rose passed the "Elon interview." As part of this preparation, the recruiter advised Rose not to ask Musk about space elevators. This struck Rose as odd, since he had never heard of a space elevator, nor been planning to ask about one. The recruiter was concerned, evidently, that a question about space elevators might trigger the boss. Musk thought that space elevators, a theoretical concept that consisted of a cable anchored at the surface of the Earth and extending far into outer space, were stupidly impractical.

Unlike most of his interviews with applicants, Musk did not press Rose with technical questions or mathematical riddles to tease out how the programmer thought. Rather, he looked at Rose's résumé, nodded, and asked if he had any questions. Rose did not ask about space elevators. But he had never spoken with someone as wealthy or accomplished as Musk before, so he didn't want to throw away the opportunity. So Rose asked about electric cars. Why was Tesla making fully electric vehicles, when hybrids made more sense by allowing a gradual transition over time from gasoline?

Musk did not answer right away. Instead, he slowly leaned back into his chair and put his hands on the desk.

"I was thinking, *My God, what have I done?*" Rose said. "He then launched into this completely passionate speech about how hybrid-electric is bullshit, and we need to skip it and go directly to fully electric cars. I don't even remember most of what he said. But I do remember distinctly thinking, *Holy crap, I want to work for this guy.* I don't understand all of the words coming out of his mouth, but he has charisma and an energy about him. If this were a war, and he said, 'Private Rose, step on that landmine,' I probably would."

Rose left the factory blown away. After receiving an offer, he moved with his wife from Bend, Oregon, to write software for rockets. Flight software is the code that goes with the rocket to space, taking over at one minute before liftoff and guiding its flight all the way to orbit. When Rose arrived in January 2009, a small team of programmers led by Chris Sloan was focused on writing code for the final flight of the Falcon 1, so he was told to start working on the Falcon 9.

Rose felt comfortable with this because of similarities between the video game industry and rockets. Rose worked in the era of consoles, when games were burned onto disks. Unlike today, when there are myriad software patches available for glitchy games, in the 2000s even a few bugs spelled death for a video game. Managing memory on console games was also incredibly important, because if a console ran out of memory, the game would crash. Similarly, if a flight computer runs out of memory, the rocket will crash.

In the video game industry, however, writing code involves shoving as much functionality into the game play as possible, to make the experience richer for gamers. Not so in rocket science. Every line of code must be closely scrutinized to ensure it works in all conceivable modes of flight. A single line of code might be analyzed and tested for weeks.

Given that, "You wanted the simplest, dumbest piece of software possible," Rose said. "We treated code as a liability."

Even so, getting a rocket to space requires a lot of code. Rose took the core software used to fly the Falcon 1 rocket and began adapting it to the larger Falcon 9. Despite a desire for the simplest code ever, the flight software for the Falcon 9 rocket would eventually run to several hundred thousand lines.

## "It's the kind of thing you do."

Like its software, the rocket's hardware was not close to being ready for flight in early 2009. Following its starring role in Capricorn One, the first stage was broken apart. The tank was shipped to Texas for additional work, and the rest of the rocket traveled back to California. There, the nine Merlins were put into a new engine section and, after additional checkouts, were sent to Texas for testing and installation on the tripod.

In July of that year, SpaceX launched its final Falcon 1 rocket. Just a few weeks later, to the shock of some employees, Musk announced the company would go all-in on the Falcon 9. It represented the future of SpaceX. Among those taken aback was Roger Carlson, who had been recently installed as site director in Kwajalein. With just his laptop and a backpack, Carlson caught the first plane from the Pacific atoll to Texas to take up his new assignment, overseeing integration and assembly of the Falcon 9 first stage. This meant sitting by the rocket for days on end, often in his customary shorts and green boots, as hundreds of engineers and technicians rotated through Texas to work on various parts of the booster.

It was often hard, hot work that summer. People had to fold into tight spaces, between engines, connecting fuel feed lines or bolting steel pins into place. Such operations often took hours, with a person confined to an area smaller than a phone booth. "We had to pass buckets inside for them to pee into in the middle of the night," Carlson said. "That's not really in the rocket scientist brochure, but it's the kind of thing you do."

The first stage had none of the critical avionics or cabling needed to actually lift off and fly to space. Engines and fuel tanks are critical, but there is much more to a rocket than just metal. Once the final "go" command is given, a rocket flies autonomously, with flight computers gathering all manner of data on speeds, pressures, and external forces such as winds. Then the rocket's flight software accounts for tiny perturbations in these variables to make sure the massive machine remains on course. A rocket, therefore, is the world's largest, most powerful flying computer.

Among the critical tasks that summer and fall was wiring and arranging hundreds of cables to carry electricity and transmit signals across the vehicle, from sensors to avionics boxes and back. Technicians also welded high-pressure gas lines along the "raceway" that ran up the rocket's exterior. Then came the software to make sure it all worked.

Rose showed up in the fall to begin testing this flight software. In particular, he wanted to ensure it could handle the precise timing needed for igniting and shutting down the Merlin engines. Due to the rocket's design, it was not advisable to shut all nine engines down simultaneously. Rather, they had to be shut down in pairs, symmetrically, otherwise the fuel lines might rupture. Rose had spent a large chunk of time coding the shutdown timing to within tens of milliseconds.

One evening, with the rocket still in the hangar in McGregor, technicians prepared to plug the vehicle into the ground support equipment for the first time. It had already been a long day, and everyone was eager to complete this task and begin flowing data back and forth from the rocket's flight computer to the ground systems. The moment was reminiscent of the scene from the movie *National Lampoon's Christmas Vacation* when Chevy Chase calls his whole family onto the front lawn to see his home lit with seasonal lights for the first time. When the rocket was plugged in, nothing happened.

Everyone looked at Rose, who had written the software. He heard exasperated sighs from people ready to call it a day. Rose thought it might be a hardware issue, but that did not stop the jokes from filtering his way through the noise in the hangar. As hours passed, people trickled out of the hangar. Rose himself stayed until well after midnight, trying to discern the problem, without luck.

The next morning, after a few restless hours of sleep, Rose returned. Instead of troubleshooting his software, he began to poke around the rocket and its interface, to ensure there was a tight connection between the rocket and ground systems. As he did so, Carlson ambled over. Rose told him he was just a software guy, but it sure looked to him like the two connecting cables were not lined up perfectly. Carlson took a closer look, and then bellowed out like a drill sergeant that no, it did not look to him like they were lined up either. They soon discovered that a slight revision had been made in the interface between its development in Hawthorne and what showed up in McGregor. So the connectors were off by a millimeter or two, enough to prevent a tight link.

SpaceX confronted a million problems like this with the first Falcon 9 as the engineers and technicians worked with new hardware developed at a rapid pace. Every day they seemed to take one step forward, and then discover a problem that set them two steps back.

Like Carlson, other engineers who had worked on the Falcon 1 decamped to Texas that summer. Zach Dunn had played a pivotal role during the latter half of the Falcon 1 program, overseeing first stage propulsion. Now the smaller rocket's launch team threw themselves into final assembly and testing of the Falcon 9 first stage. Dunn had worked exceptionally hard on Falcon 1, sleeping in the company's factory for many consecutive nights. But the Falcon 9's debut flight campaign proved still more rigorous as the rocket moved from the hangar to the tripod.

"I think that time in Texas was probably the hardest I ever worked, period," he said. "We worked seven days a week for months on end getting the first stage tested."

The day would start around nine in the morning, although there were no set arrival hours. Work continued until well after dark, and often for twelve hours or longer. When engineers were in McGregor they did not really have any weekends off, even if their stints stretched for a month or longer. You went to Texas to work. Occasionally the local workers would take a weekend day off during that never-ending summer. The hit single by Green Day, "Wake Me Up When September Ends," became the soundtrack of the season.

And they weren't exactly working in the most comfortable environment. The surrounding terrain was rural enough to require safety warnings about critters like rattlesnakes. To avoid snakebites, workers were advised to watch where they put their hands and to always wear gloves. In that part of the country, crickets hatch during the spring, mature during the long summer months, and then die off in the fall. These noisy infestations are drawn to bright lights, congregating around the entryways to buildings. In McGregor, the insects especially liked the blockhouse, where tired and sweaty engineers would gather to conduct tests. As SpaceX put the flight-ready version of the Falcon 9 rocket on the tripod that autumn, a plague of crickets built up outside the blockhouse, blowing up like snow drifts. As they died, they added their mephitic odor to the interior of the blockhouse, where trash piled up everywhere. "It was an absolute pigsty," Dunn said.

The elements in Central Texas also proved to be occasionally uncooperative amid the unrelenting toil. As summer turned to fall, periodic cold fronts and squall lines would slam into engineers and technicians toiling a hundred feet in the sky on the tripod. In the morning they would blast Metallica's "For Whom the Bell Tolls" to psych themselves up for another long day, starting by reviewing data from the previous day's tests and plotting their activities for the day and night ahead. Musk kept the pressure on, constantly telling the teams they needed to complete the first stage within two weeks. It was never soon enough for Musk's taste, but finally, by mid-November 2009, the rocket was deemed ready to ship back to Florida for final launch preparations.

This presented a whole host of new problems, however. SpaceX had never tried to move something that large, an entire first stage, halfway across the country. Although the drive from Central Texas to Florida's Space Coast can be made in about sixteen hours in light traffic, the Falcon 9 rocket's first stage would not enjoy traveling the wide open freeway.

Rather, it would make the road trip from hell.

## Free Candy

Laid horizontally on a flatbed trailer, the Falcon 9 rocket exceeds the dimensions of a semitrailer in every direction. A normal long-haul truck is eight feet wide. The large, steel cradle carrying the Falcon 9 first stage was fourteen feet in width—nearly twice as wide as a standard semi. The top of a semi truck's trailer typically stands thirteen and a half feet above the road, fitting comfortably under freeway overpasses, which are commonly fourteen to sixteen feet tall. Due to the way it sat on the cradle, the first stage topped out at seventeen feet, eight inches above the ground. Finally, a semitruck, combining both the tractor and the trailer, is about seventy feet long. The front of the truck to the back of the Falcon 9, which had a large overhang, measured 120 feet. This was therefore not just a wide load, but an extraordinarily tall and long load.

A commercial driver simply could not pull onto Interstate 10 with the rocket and drive from Texas to Florida.

NASA and other U.S. rocket makers typically moved large boosters by air or sea. NASA extensively modified two Boeing 747 airliners in the 1970s to fly the space shuttle orbiters across the country. The massive Saturn V and shuttle tanks were moved by barge from Michoud Assembly Facility in Louisiana across the Gulf of Mexico to the Kennedy Space Center. But such modes of transport were slow and expensive, and Musk had sized the Falcon 9 rocket just small enough to be moved along America's roads. Theoretically at least.

SpaceX hired a trucking company to move the completed stage, which plotted a route for two commercial drivers to follow roads designated for wide loads. The first day went smoothly, but the trip bogged down once the semitrailer reached Louisiana. After starting in the northern part of the state, the route ran south of Interstate 10, along bayous and marshland, often down backcountry roads. While these roads did not have overpasses that would obstruct the truck's safe passage, there were plenty of low-hanging traffic lights and power lines. This necessitated a handful of escorts, including Carlson.

He often rode with Chris Thompson, who'd been a founding employee alongside Mueller, inside a Ford Flex SUV. Another vehicle in the little caravan had a pole on its bumper, with a flag topping out at seventeen feet, eight inches. If this flag ran into an obstruction, Carlson and Thompson would hop out of the Ford Flex and use large sticks to lift the power line or other obstacle until the rocket passed beneath. Then, running back to their SUV, Thompson and Carlson would veer off road to pass the rocket to resume their position.

"Every morning we remeasured the pole on the front car," Carlson said. "One day we found that it had slipped and telescoped down a few inches. That was terrifying."

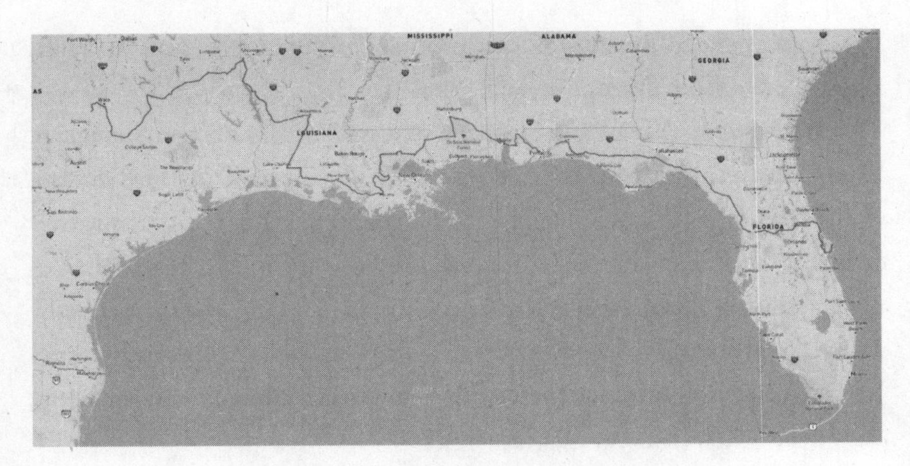


The tortuous route used for the F9 road trip in 2009. | PHOTO CREDIT: ROGER CARLSON

To facilitate the journey across the back roads of Louisiana, SpaceX hired off-duty state troopers when they could. This allowed the caravan to avoid stopping at red lights and more smoothly pass through intersections diagonally when there was no other way for the exceptionally long vehicle to make a turn. One day, during the journey in Louisiana, the truck had to negotiate an intersection where two highways met, with multiple large, low-hanging traffic lights. As the rocket painstakingly progressed forward, traffic began to back up for more than a mile in each direction. The SpaceXers were frazzled as they hurried, tense about the traffic. But their hired-gun state trooper took it in stride. "Let 'em wait," he said, looking in one direction, and then the other. "If I wasn't here with you guys, I'd be sitting somewhere on an off-ramp, writing speeding tickets. Take your time and do what you need to do."

At night they had a Sunseeker RV to sleep in when there was no hotel nearby. One of the popular internet memes at the time depicted a creepy van with "Free Candy" painted on the side. This was dark humor, playing off the "strangers with candy" warning parents give to children.

As a joke, before the caravan departed from McGregor in Central Texas, site director Tom Markusic had painted "Free Candy" on the side of the motorhome. After Carlson snapped a photo of Thompson emerging from the van, the little crew started to tease Thompson about getting arrested. One of the state troopers, who stood about five feet, six inches tall, told the taller Thompson that he could cuff him with ease.

"No way," Thompson replied.

The trooper gave him a "watch me" look and then, in a single move, trussed Thompson like a perp. "The next thing you know I'm leaning against the car, in cuffs," Thompson said.

On Sunday, November 22, the route through Louisiana called for a 90-degree turn, which was difficult for the 120-foot truck to swing through. When the truck reached the intersection, the turn was even tighter than anticipated. To make room, Carlson and the other SpaceX-ers exited their vehicles and started to cut down road signs with hacksaws. They had already used these saws many times before, hacking at trees and other obstructions. As they worked, Carlson, Thompson, and the others grew desperate. Once again cars were backed up for as far as they could see. Their trooper escort warned that by late afternoon in this part of southern Louisiana, everyone would be getting drunk, watching the New Orleans Saints football game. During the drive home, the locals were bound to get angry having to wait and might exercise their frustrations on the rocket.

Faced with a no-win situation, the SpaceX team finally decided to go straight through, off the route, to clear the intersection. By then, after a long, nerve-wracking day, everyone needed a break. The caravan began to separate, with individual vehicles looking for somewhere to park the rocket overnight. One driver found a large warehouse with a loading dock and called the emergency number on the door. For $200, the manager said, SpaceX could park, but he needed a damage waiver. The team slept outside, picketed around the rocket protectively, wondering what might go wrong the next day.

They would soon find out. As their Louisiana adventures continued, the SpaceX caravan pulled into a small town square straight from the set of the movie *Back to the Future*, complete with a towering church. As the main road entered the square, the rocket had to navigate a sequence of 90-degree turns. The procession moved slowly, with Carlson and the other escorts walking beside the rocket, communicating over simple CB radios. The shops and storefronts were built nearly to the edge of the road. Sure enough, as the truck began to make one turn, the end of the rocket appeared on course to strike an overhanging roof.

Carlson and the others immediately pressed the "talk" button on their CB radios to warn the truck driver. But with that type of radio, only one person could talk at a time, so their screams all canceled one another out. Due to the delayed warning, the truck driver did not stop in time. The flange, or rim of the top of the rocket, swung around and crashed into a wooden building, damaging both the rocket and the structure. For Carlson, Thompson, and the other SpaceXers (Scott Moon, Brad Obrocto, and Joe Fitzgerald), it was a gut-wrenching moment. They had been working long days, at the edge of their wits. The truck driver was in tears. The accident broke them.

"There were only like five of us on the road, and it was awful," Carlson said. "Everyone was yelling and screaming. Everybody just kind of melted down. It was the worst moment in my life."

They all stood there for a few minutes, gathering themselves. Then, there was nothing to do but press onward. The driver clambered into the semi and backed up. As he did so the rocket's flange could be heard grinding back through the building until it was free. Thompson, the senior official on the trip, made the call to Musk. Well knowing the bitter reaction he could expect from such an accident, Thompson steeled himself. But when he explained the metallic flange had rebounded into place, and there were only minor scuffs in the paint, Musk shrugged it off. He thought the tank would be fine and urged the transport team to keep moving.

Musk's mood changed the next day when Carlson submitted his daily update to the company. Carlson took pride in these reports, widely shared back at SpaceX headquarters. They described technical progress made during the previous day's travel, and Carlson often added a bit of color commentary. As part of his report on the rocket bumping into a building in Louisiana, still reeling from the experience, Carlson compared it to the third flight of the Falcon 1 rocket in 2008, when the first and second stages collided, dooming the mission. Not everyone back in Hawthorne appreciated the tenor of that update, however. At least one employee complained to Musk about its flippancy, suggesting Carlson was not taking his responsibility seriously.

This set Musk off. He telephoned Thompson, then driving the Ford Flex down yet another Louisiana byway. After berating Thompson, Musk ordered him to summarily fire Carlson. The physicist was sitting next to Thomspon, oblivious to the conversation. Thompson replied that he was not Carlson's supervisor and did not have the responsibility. Musk said that if Thompson did not fire Carlson, he would fire Thompson first, and Carlson second. The call ended. Thompson turned to Carlson and nonchalantly said everything was fine. They drove on down the road. Musk never raised the issue again.

Eventually, the caravan crossed the Pearl River, finding fewer sharp turns in Mississippi, Alabama, and Florida. They never stopped once reaching Florida, because the state allowed oversized loads to travel on highways at night. And there was no time to lose: the rocket faced a hard deadline at 5 PM on Tuesday, November 24, when Florida's roads would close to large loads for nearly a week to accommodate Thanksgiving travel. The caravan pulled into SpaceX's hangar at 3:21 PM on Tuesday, having driven straight through the night.

It had taken ten exhausting days to make the 1,200-mile trip.

Among those waiting for the Texas crew was Catriona Chambers, the avionics engineer who had spent most of the year working on the rocket's first stage in McGregor. Some things from that experience, such as the smell of dead crickets, she did not miss.

"When they rolled in, the stench of McGregor rolled in with them," she said. "We knew McGregor had arrived. One of the first things we did was vacuum the crickets out of the stage."

It had been a long, exhausting journey from Texas, traveling glacially at an average speed of just 10 mph. But the trip taught SpaceX some hard-won lessons. The company's engineers developed a customized transporter that placed the rocket much nearer to the road, in a special cradle with a narrower base. This allowed Falcon stages to be transported on freeways, at a steep discount to traditional rockets.

## Work like monsters, party like rock stars

The arrival of the first stage in Florida put SpaceX into the homestretch for the big rocket's debut launch. The second stage arrived on January 26, 2010, and the vehicles were soon mated into a single rocket. One month later, SpaceX completed a critical fueling test, and by mid-March it had successfully fired the vehicle's engines during a full-power static fire test. The rocket and ground systems were largely ready to go.

However, Air Force officials were unyielding on a key safety issue: the capability to destroy an errant rocket. This is known as a flight termination system and, in essence, involves strapping a bomb to the first and second stages of a rocket, and connecting each of them to an igniter and a receiver. If a rocket were to stray off course, a range safety officer would send a signal to the igniter, blowing up the rocket before it endangered anyone on the ground.

In Musk's mind, this was a simple operation. "This flight termination system is just like a light bulb, right?" he had said to Helms in 2008. "You turn it on, or you turn it off. And if you turn it on, the rocket blows up." Well, yes. But there is a lot of complexity involved in making sure this particular light bulb does turn on.

This unenviable task eventually fell into the lap of Hans Koenigsmann, the fourth employee at SpaceX. Originally, the German engineer had managed avionics for the Falcon 1 rocket, but over time his role evolved to focusing on flight safety. Launching from Omelek Island on Kwajalein Atoll in the Pacific posed fewer risks. The tiny island launch site was surrounded by water, and there were no major population centers. As a result, no explosives were needed to end the flight of the Falcon 1. The engine's thrust would simply be terminated. But in Florida valuable aerospace properties crowded the Falcon 9 launch site, and the cities of Cape Canaveral and Cocoa Beach were nearby. So a wayward rocket's flight might need to be ended rapidly with a detonation.

Koenigsmann turned his full attention toward the Falcon 9 rocket's flight termination system at the end of 2009. All of the hardware that would fly as part of this system had to be "qualified," which meant undergoing an elaborate series of tests. The Air Force literally had a big book of requirements for such tests and defined the process line by line. "It was just really scripted, and very stiff, in how you qualify your hardware," Koenigsmann said.

Facing these strict requirements and concerned about the timeline for getting a flight termination system approved, Musk opened up SpaceX's checkbook. Koenigsmann bought two space-certified receivers for $250,000 to receive the signal from the ground and transmit it to the ordnance onboard the rocket.

The biggest problem came with the batteries to power the flight termination system, including the receiver and the firing circuit. This power source had to be independent of the power onboard the rocket. Koenigsmann decided to use the same batteries that had been on the Falcon 1 rocket, each about the size of a hardcover nonfiction book. The Air Force specified that the batteries must survive twenty-four cycles between –20 degrees and 160 degrees Fahrenheit (–29 degrees and 71 degrees Celsius), without any recharging.

"They were really extreme with their environmental tests, which had nothing to do with the real environment," Koenigsmann said. "And batteries don't like thermal cycles." During the course of this testing, if a battery violated the voltage requirement, even briefly by dropping down to 18 volts instead of 20, it would fail the entire qualification test. And some did.

Over the course of several weeks, it got to the point where SpaceX engineers were using a CT scanner to investigate individual battery cells to ensure there were no glitches. By early 2010 Koenigsmann and the flight termination system were holding up the debut launch. The German engineer and his small team brought in people from other areas of the company, as well as interns, to search for defective cells and repeat thermal tests. It was a desperate, twenty-four-hours-a-day, nights-and-weekends effort. During this process Koenigsmann was managing up as well as down. Musk wanted to be informed frequently on progress.

"I tried to be as frank as I could with Elon," Koenigsmann said. "But it could be painful. He does not speed things up by yelling at people, in my eyes. This creates anxiety and unnecessary stress. People were stressed out anyways. It would have helped them more if he just, you know, got them some ice cream or something that was nice, rather than being yelled at."

Then splitting his time at Tesla, Musk felt his work at the electric car company made him an expert on batteries, although the flight termination system battery chemistry was completely different than that used in automobiles. He was upset with Koenigsmann, with the Air Force range officials, and with the Federal Aviation Administration, which would be providing the launch license.

Koenigsmann and his team pressed on, and the arrival of a new commander of the 45th Space Wing in February helped the process move forward. Major General Ed Wilson urged Schindzielorz and his range officials to work with SpaceX to certify the batteries and the rest of the flight termination system.

By the spring of 2010 the Falcon 9 team really started to gel, forging itself through the crucible of hard work and late-night shenanigans. For the most part they were young, in their twenties, with only a few adult "chaperones" like Koenigsmann and Buzza. The engineers lived in hotels and motels, or in extended-stay housing and apartments, and when they weren't working they played rock and roll at all hours of the night. One of the closest watering holes to the main entrance of Cape Canaveral was Fishlips Waterfront Bar & Grill, and this became a popular place to blow off steam. Fishlips considered itself a family-friendly restaurant, so it was not uncommon for the SpaceXers to get kicked out when their antics approached *Animal House*–level rowdiness. The chaperones would introduce themselves to the manager and say that if there was a SpaceX problem to please call them, rather than the police.

"We worked like monsters," Zach Dunn said. "But we also partied super hard."

Sometimes bounced partiers would try to climb back over the fence to rejoin friends at Fishlips. Others got more creative. Then in his late twenties, Dunn had long hair when he got kicked out one night, told not to return. His solution was to get a haircut the next day, radically shortening his locks. When Fishlips was not accommodating, the SpaceXers walked a block over to Grills Seafood Deck & Tiki Bar. There, one time, a drunken Dunn ate several rock shrimp that still had their shells on without realizing it. In response to this, the waitress quipped, "You have to be smarter than the food you eat."

Nevertheless, these were the people who were going to revolutionize spaceflight.

By planning to launch a rocket from Florida rather than from the middle of nowhere in the Pacific Ocean, SpaceX had jumped into the big leagues. On their way to work, the engineers and technicians would drive by historic launch pads where Alan Shepard and Gus Grissom had made America's first human flights into space. They had front-row seats for some of the final space shuttle launches and daydreamed about one day maybe launching some of those astronauts into space themselves.

The Sunshine State offered a pleasant contrast to chill air and drab grays where Catriona Chambers grew up, in the Scottish Highlands city of Inverness. She studied engineering at Oxford University and followed her brother Andrew to SpaceX in 2005 to work on avionics. In Florida, the company rented her a beachfront condo for nearly a year ahead of the first launch. Chambers lived with one of her best friends at SpaceX, Florence Li. To break up the long days, they would find time to go jogging on the Air Force station beach, passing by old launch towers in the hazy distance. If she needed a short nap, Chambers would slip inside the Falcon 9 interstage for a few minutes, where acoustical blankets deadened noise from the outside world. Late at night she would call her husband back in California, falling asleep talking to him.

Another difference between Kwajalein and Cape Canaveral was time; the launch team had a little more of it to savor the experience. The final days and hours before a Kwaj launch had been so hectic that there just weren't many moments for reflection. Koenigsmann and Buzza wanted a better experience for the Cape launches. So on the evening of June 2, two days before the planned launch date, they invited dozens of the Falcon 9 engineers over to their apartment. As the crew gathered in the living room, Koenigsmann played a DVD titled *Liftoff: Success and Failure on the Launch Pad*. Watching footage of past rocket disasters and explosions helped the launch team clear their minds and realize that if the worst happened it would not be the end of the world.

## The Falcon 9 takes flight

On the afternoon of the next day, less than twenty-four hours before the planned launch, a classic summertime thunderstorm rolled off the Atlantic Ocean. The heavens opened up, and torrential rainfall doused Cape Canaveral with as much as three inches of rain in a single hour. Four tall lightning towers protected the rocket from electrical strikes, but there was no defense against the nearly sideways rainfall striking the pad. After the storm passed, engineers resumed checkouts of the vehicle. Pretty quickly, they discovered a problem with the second stage. The stream of data coming from the stage was intermittent, like a mobile phone call dropping out. Water must have intruded into the antenna compartment.

This necessitated a trip to the launch pad to investigate. There, Musk huddled with some of his engineers, including Buzza as well as avionics engineer Bulent Altan. The safe option was to remove the cylindrical Haigh-Farr telemetry antennas and install new ones. But that would have delayed the launch considerably because the antennas were hardwired into the stage. Instead Musk and his team tried a different approach. They scrounged up a heat gun, which is a handheld device similar to a hair dryer, and put Altan into a lift. For several minutes, he waved the gun back and forth across the affected area to evaporate moisture from the antenna compartment.

As Altan and others scurried to fix the rocket that evening, Musk wandered around asking anyone at hand—technicians, junior engineers, and company vice presidents alike—the same question: "What can we do to go faster?" Initially the launch team thought he meant the frantic effort to dry out the second-stage antennas. But actually, Musk was referring to future operations and his goal of reaching a rapid launch tempo. Not receiving a sufficiently satisfactory response, Musk eventually sat down on the tarmac, next to the horizontal rocket. Sitting with his legs crossed, Musk rested his chin on tented hands and stared into the unknowable future, as if he were meditating for answers. A few hours later, on the drive back to his hotel, Musk was still contemplating the need to streamline Falcon 9 operations and other future problems. Well after midnight, as Buzza drove, Musk asked his launch director about increasing production, landing Falcon 9 cores, and even the as-yet-undesigned Falcon Heavy.

Buzza had more pressing concerns. After grabbing a short nap, he and other members of his team filed into the windowless launch control center around dawn and continued checking the rocket and its systems. While the stream of telemetry data marginally improved overnight after Altan's intervention, it still glitched out from time to time. Most of the launch team felt this would be a nonstarter, fearing losing contact with the second stage in flight. Less than an hour before the launch window opened, Musk conducted a poll to assess the readiness of the rocket. During this poll Buzza, Altan, and the company's vice president of avionics, Jeff Ward, all spoke at some length about their concerns with the spotty second stage telemetry. Effectively, but not explicitly, they were saying they were "no go" for the launch.

But then Musk shocked them all. "Well, okay," he said after completing the poll. "I'm not hearing any reason why we shouldn't go."

Accordingly, the launch team began counting down toward a 1:30 PM liftoff, local time. The count proceeded smoothly until 1 PM, when a wayward boat intruded into the designated launch area offshore. The range safety officer establishes such no-go zones for boats, so they won't get hit by falling debris if a rocket breaks apart during ascent. In the launch control center, the engineers tried to keep the mood light. Someone back at the company's headquarters photoshopped a picture of Richard Shelby on a jet ski in the restricted zone. Shelby was a powerful Republican senator from Alabama who represented the interests of United Launch Alliance, which had a large manufacturing facility in his state. The photo was passed around the control room.

Soon enough, SpaceX got word that help was on the way. The new Air Force commander, Wilson, had dispatched a Black Hawk helicopter to intimidate the boat with a blast from its propeller blades. He then telephoned Buzza and told him the restricted area would be cleared one way or the other. The boat moved on, and so did the countdown.

The clock ticked down to the final seconds, but then a sensor reported that not enough TEA-TEB igniter fluid was flowing into one of the rocket's nine Merlin engines. This triggered an automatic abort and put Kevin Miller on the hot seat. Mueller came over to see Miller first, and Musk soon followed. Over his headset, Miller had Zach Dunn in one ear and Buzza in the other, also asking for updates. The launch window only extended to 3 PM local time, so Miller felt pressured to work the problem quickly.

Not getting enough TEA-TEB into an engine, if it attempted to ignite, would be bad. During one test in McGregor, the flow of igniter fluid into the engine failed and dramatically blew up the engine's gas generator. For the debut launch of the Falcon 9, SpaceX had set conservative boundaries on the acceptable level of TEA-TEB flow, so Miller got buy-off from Musk, Mueller, and the others to allow more generous limits. After updating the rocket's flight software to accept the new parameter, liftoff was reset for 2:45 PM.

This time the countdown was clean. Not a single engine aborted. And then, shuddering and shaking, the rocket cleared the lightning towers and rose steadily into the heavens. As it broke through the sound barrier, the Falcon 9 produced a visible shock wave. Climbing farther upward into the thinning atmosphere, its plume of exhaust expanded rapidly.

Musk nervously followed liftoff from inside the control room, standing behind Chambers's desk, watching a large video monitor above her station. They both could feel the rumble of the rocket as its furious sound waves swept across the building. It felt incredible to not just see their rocket climbing, but to feel it as well. Chambers watched the data on her screens and tried to play it cool. Internally she wanted to leap out of her seat in joy. After the first stage completed its burn, she could resist no longer, jumping to her feet and punching into the air.

After the nine engines completed their task, Musk walked away from the video screen full of giddy satisfaction. Whatever happened next, this was good enough. He could sell the nominal first stage performance as a successful mission.

Just outside the control center, Roger Carlson watched alongside hundreds of employees and their family members, standing on the lawn and looking up expectantly. After driving this Falcon 9 first stage across the country, Carlson had monitored the attachment of the first and second stages in the Florida hangar. He observed the painstaking work that had gone into installing the mechanism that would push the stages apart after all of the propellant in the first stage was consumed. It had taken three days to install the stage separation system, and he knew how many ways it could fail. He stood beside another SpaceX engineer, David Freidhoff, who bore some responsibility for the separation system.

In the final seconds of the countdown, Freidhoff had turned to Carlson and said, "Man, I hope that thing works."

With the rocket climbing into the sky, the SpaceXers on the lawn whooped and hollered and felt jubilation. But Carlson and Freidhoff held back. At 182 seconds into the flight, the rocket's main engines shut down. Four seconds later, the separation mechanism triggered. As the first stage fell away, the single Merlin engine on the second stage ignited. It began flying. Then, and only then, did Carlson and Freidhoff join the increasingly frenzied celebration.

The Falcon 9 rocket would go all the way to orbit that day. Its second stage fired for several minutes, reaching low-Earth orbit with a "boilerplate" version of the Dragon capsule, meant to simulate the mass and shape of the actual spacecraft. The Falcon 1 had failed on its first three tries, but now SpaceX had nailed orbit on the first time out with its much larger and more complex Falcon 9. The rocket had flown with astonishing accuracy into its target orbit.

After their rocket did its job, the launch team emerged into the bright sunshine, squinting and rubbing their eyes, to reconnect with their friends. They had spent long days, for weeks on end, tucked inside that control room with no windows, training for what they had just accomplished. "We felt so triumphant, and it showed in our stride," Chambers said. "We had family and friends who had been right outside for hours. The feeling was amazing."

It was a stunning achievement, though not flawless. In the first two seconds after liftoff the rocket rotated 60 degrees. The engines had been slightly misaligned, causing this turn. However, the rocket's guidance, navigation, and control system sensed the rolling motion and stopped it before the Falcon 9 spun out of control. The second blip came much later. SpaceX intended to relight the second stage Merlin engine to make a controlled reentry into the Pacific Ocean. It is standard procedure to preserve a small amount of fuel for this purpose, because it prevents the upper stage from flying in low-Earth orbit as a large piece of uncontrolled debris. And any metal that doesn't burn up in the atmosphere will fall harmlessly into the ocean. Launch companies typically aim for Point Nemo, 48 degrees South, the spot in the Pacific Ocean farthest from land. Nemo is surrounded by 1,000 miles of ocean in every direction.

However, before reentry into Earth's atmosphere, SpaceX lost roll control of the Falcon 9's upper stage. When they emptied the propellant tanks prior to its return, the stage was spinning as it spewed propellant. Below, it was a few hours before dawn in New South Wales. As early rising Australians looked up, they saw a bright, swirling lollipop. Local TV and radio stations were inundated with UFO sightings as the stage burnt up in the atmosphere.

The Falcon 9 took flight at a pivotal moment in U.S. space policy history. A battle was raging between the White House and Congress over the future of human spaceflight. As the space shuttle's retirement loomed, the big aerospace companies holding contracts that supported it pushed Congress to receive similarly lucrative funding for a new generation of rockets. But the Obama White House wanted to press pause and give emerging players like SpaceX a chance to bring down spaceflight costs. Thus, the first Falcon 9 launch offered a referendum on President Obama's space policy. If the rocket failed, naysayers would be justified in their view that commercial space was not ready for prime time.

Obama's principal advisor on space policy, NASA Deputy Administrator Lori Garver, felt this all too keenly. "I was well aware that not only my own reputation, but the success or failure of the Obama Administration's space policy, would be largely determined by the outcome of the SpaceX launch," she said.

Though the Falcon 9 soared, its success hardly mollified Musk's critics or those in Congress aligned with the traditional space contractors. This should have been a great moment for the country. SpaceX's rockets were made entirely in America, giving the nation a strategic advantage. At the time, most of the country's national security satellites reached orbit on an Atlas V, for which United Launch Alliance had to buy engines from Russia. The Falcon 9 promised affordable launch independence.

Alas, no. Space policy leaders on Capitol Hill such as Shelby, allied as they were with the existing aerospace power brokers, offered dismissive responses. Even the senior U.S. senator from Texas, Kay Bailey Hutchison, cast the launch in negative terms. "Make no mistake, even this modest success is more than a year behind schedule, and the project deadlines of other private space companies continue to slip as well," she said. This reaction seems almost incomprehensible. Hutchison was a pro-business Republican, and SpaceX an entrepreneurial success. The company had a growing presence in Texas with nearly 100 employees at the McGregor test facility. Moreover, SpaceX's success was critical for the space station. The NASA field center that managed the station, in Houston, was located just a few miles from where Hutchison grew up.

But for all of its raw physical power, the Falcon 9 could not break the big aerospace lobby.

Regardless, Musk knew his company had done something special. His people had worked very, very hard to get this rocket off the pad. As they pushed through the last steps, he'd promised his employees a break from the nonstop crunch mode. True to his word, following the successful launch of the Falcon 9, Musk said the whole company would take the week of July Fourth off, and almost everyone did.

"You didn't get any emails for a week," Chambers said. "The whole company just took off to places around the world. It was the best thing ever." She flew to Bolivia to climb Huayna Potosí mountain, which tops out just shy of 20,000 feet.

Robert Rose, whose software helped steer the Falcon 9 into orbit, did not go anywhere. He had waited in Florida as long as he could, but he returned to California three days before the launch to be with his wife for the birth of their second son. "I didn't get to see the birth of the Falcon 9, but I did get to see the birth of my child," he said. Rose spent the week off with his young family.

Buzza, the triumphant launch director, rented a recreational vehicle to travel around California with his wife Jo, daughters Abby and Brandy, and his mother-in-law. They went horseback riding in Mammoth Lakes and celebrated the holiday with fireworks in Lake Tahoe. After sacrificing so much time away from his family during the Falcon 9 launch campaign, Buzza deeply appreciated the time off, grateful to be with his loved ones at an important time. Jo's mother was near the end of her life and said that more than anything she wanted to take a shower on the beach, right next to the surf. And so she did one evening, on Pismo Beach, along California's Central Coast.

She died two weeks later.

| 4 |

# FLIGHT TWO: DRAGON'S DEBUT

*September 2010*
Hawthorne, California

A few months after the successful debut of the Falcon 9, Musk summoned Chris Thompson to his cubicle in Hawthorne. The leader of SpaceX had a workspace about twice as large as most of the company's employees, but it was egalitarian in that it had no walls. From his perch on the first floor, near the Hawthorne factory's main entrance, Musk could see the majority of SpaceX's engineering team at work. And they could see the boss.

Thompson and Musk did not have the easiest of relationships. Eight years after the founding of SpaceX, all of the company's original vice presidents remained, having survived the difficult Falcon 1 years and sharing in its ultimate success. But more than the others, Thompson stood up to Musk, and these regular clashes strained their relationship.

Thompson would be the first of the vice presidents to leave permanently, fewer than two years later. Now, however, there was no clash brewing. Musk wanted to talk about what the Dragon spacecraft should carry on its first spaceflight.

"I could tell right away that he was trying to maintain a straight face," Thompson said. "He was in poker face mode, and it seemed like he was going to lose it at some point."

Dragon's payload, Musk told Thompson, was going to be "top secret." Thompson could not fathom why SpaceX would fly a top-secret cargo inside its own spacecraft. This was a straightforward demonstration flight for NASA, after all, not a spy mission for the military. When he asked how secret the payload would be, Musk responded that it was, actually, "super top secret." Musk kept bantering away, emulating the British comedy troupe Monty Python in his repetitive responses to Thompson's increasingly baffled questions.

Finally, he spilled the beans. Dragon would carry a twenty-five-pound wheel of Le Brouère cheese into space.

"When he told me it was going to be a wheel of cheese, I lost my shit," Thompson said. "I was literally in tears. And Elon, he lets out one of the horse laughs he does, where he's laughing so loud it sounds like somebody's coughing up a lung. And he's got tears coming out of his eyes, too. It's that funny. And I'm thinking to myself, *Oh my God, we're going to fly a frickin' wheel of cheese*." Then for grins they watched the Monty Python "Cheese Shop" sketch, which had inspired the gag, on Musk's computer.

Flying a block of cheese to honor a forty-year-old comedy sketch may seem a bit odd, but Musk had his reasons. First, he does like to have fun, and he thought this was a great gag. He also knew that flying a wheel of cheese would be irresistible to media covering the flight and, if not revealed until after the launch, would make for a good second-day story. And finally, this differentiated SpaceX from the buttoned-up aerospace contractors who launched rockets for NASA and the U.S. Department of Defense; like a mullet, SpaceX was business in the front, and party in the back.

For Thompson, this was a special moment, one of the times he truly enjoyed working with Musk. He took the gag and ran with it, swearing a few members of his structures team to secrecy and designing a food-grade silicone seal to prevent the cheese from leaking or outgassing during flight. Then, playing on Musk's idea, he took a snippet from the *Top Secret!* movie poster, of the cow in boots, and affixed this as a label on top of the cheese wheel. SpaceX would also fly hundreds of patches inside Dragon, each with a number on it. After the flight, each employee would be given the patch that matched their employee number. Thompson, when he left SpaceX in April 2012, would proudly take home patch number two.

## The Magic Dragon

Dragon marked an important step toward Musk's ultimate goal: settling Mars. Rockets come in handy for breaking the chains of Earth's gravity, but a different type of vehicle is necessary to fly among the stars. A rocket's flight typically lasts less than ten minutes, but a spacecraft must operate for months or years as it flies to destinations in space. Soon after SpaceX started to build rockets, therefore, Musk pushed his company to start thinking about spacecraft construction. To keep this vision before his employees, Musk commissioned construction of a spacecraft mock-up, complete with crew seats, shortly after he founded SpaceX. He named the vehicle "Magic Dragon," a clear allusion to the folk song "Puff the Magic Dragon" and its thinly veiled references to smoking marijuana.

More akin to a prop than a real spacecraft, the small capsule mock-up lacked the flash and dash Musk liked his products to convey. SpaceX wound up storing Magic Dragon under a big blanket, only occasionally showing it to VIPs visiting the company's first, small factory in El Segundo. As SpaceX began to vie for government contracts, Musk realized the Magic Dragon's 420ish name might not offer the best message to a family-friendly agency like NASA. So he dropped "Magic" from the spacecraft's appellation, and the company started referring to it simply as Dragon.

Focusing on a spacecraft early on proved a smart move for SpaceX and Musk. In 2006 and 2008, NASA threw the fledgling company two lifelines because of Dragon. First, SpaceX won an initial contract in 2006 to start working on a means of delivering food, water, scientific experiments, and other supplies to the International Space Station. This $278 million award was instrumental in the company's early growth and financially helped SpaceX survive three failures of the Falcon 1 rocket. While this represented a large amount of money for the company of a few hundred employees, SpaceX still had to build the Falcon 9 rocket, the pad at Space Launch Complex 40, and, of course, Dragon. This work proceeded even as much of SpaceX focused on getting the Falcon 1 into orbit in 2007 and 2008.

"It was an insane amount of money for SpaceX at the time," said David Giger, a young engineer tapped to lead propulsion systems for Dragon. "But it was also a two-edged sword. There was a huge sigh of relief in the sense that wow, we won this big contract with so much revenue. But then it was like, oh my God, how do we do this when we're trying to fly the Falcon 1 at the same time?"

Giger earned the job of spearheading Dragon's thruster development almost by default. In those days the company's primary goal was getting the Falcon 1 flying, and everyone with engine experience worked on the

Merlin 1C. When Gwynne Shotwell needed technical help writing a proposal for NASA in the spring of 2006, the most recent propulsion hire, Giger, was tapped. He joined the Dragon team, led by Steve Davis, which started by asking basic questions. What sort of thermal protection should Dragon use for its fiery reentry through Earth's atmosphere? How many different kinds of engines should there be? How should the cabin interior be laid out?

Two principles drove the team's decisions: cost and simplicity. Musk directed the team to use just one kind of engine for all of Dragon's functions. Previously, most of the orbital spacecraft built for cargo and crew had used multiple sizes of engines to each perform a specified task. For Dragon, a new thruster had to be designed that could both fire a sustained burn for as long as twenty-five minutes, and also operate in short bursts that lasted for a tiny fraction of a second. Spacecraft engine expert Dean Ono led this design work, and these became the Draco engines, as *draco* is a Latin word for dragon.

Small though the Dragon team was, they were the key to SpaceX's survival in 2008. Musk kept his financial situation fairly close to the vest, but it was dire that summer. With both SpaceX and Tesla flailing, he was weeks away from personal bankruptcy. By the fall of that year rumors of SpaceX not being able to meet payroll were actively percolating among the company's hundreds of employees. Salvation came in December 2008, when SpaceX landed a second contract from NASA, called Commercial Resupply Services. This required Dragon to fly a dozen resupply missions to the space station. Over the lifetime of the contract, SpaceX would be expected to deliver the equivalent mass of about ten small elephants to orbit. In return, SpaceX received far more funding than it had raised to date, $1.6 billion.

It was big money for a big job. No private company had ever built a spacecraft that visited the International Space Station. Only governments had. And for SpaceX to stay true to Musk's vision of reusable spaceflight, Dragon could not just fly to the space station, have astronauts offload its cargo, and then fall back into Earth's atmosphere and burn up like other spacecraft. Instead, Dragon had to return to the planet in one piece.

SpaceX aimed to recover Dragon by splashing it into the Pacific Ocean and plucking it from the water. The idea was not new. NASA had returned its Mercury, Gemini, and Apollo spacecraft to Earth in the ocean. But the last of these missions had flown more than three decades earlier, and NASA had been able to call on the U.S. Navy for help in recovery.

Lacking a navy, SpaceX called on Roger Carlson and a small band of about a dozen engineers and technicians. On its debut flight Dragon would make nearly two orbits of the planet before splashing down about 400 miles off the coast of California. What SpaceX engineers did not really know, however, is what condition the spacecraft would return in. It might be charred to a crisp or leaking hazardous chemicals. Carlson and his small team would have to replicate what a Navy aircraft carrier, a host of support ships, and several helicopters had done in the 1960s and 1970s.

First, Carlson set about finding a fleet. Unlike the U.S. Gulf Coast, where oil companies operate hundreds of offshore platforms, with all manner of work boats available, Southern California had little of this industry. Unable to find a suitable boat with a large A-frame crane to lift Dragon from the ocean, Carlson decided to rent a barge and modify it. He selected a boat 245 feet long because, if Dragon were found to be hazardous, he wanted to have room on the deck to spread out. SpaceX rented a mobile crane and chained it down at one end of the barge, and then welded those chains to the deck. Carlson's team also outfitted a shipping container to serve as a Dragon control room on the deck of the barge and built a rubberized "nest" to set the vehicle in. Mindful of hazardous contingencies, the team installed nozzles around the nest to spray Dragon with sea water if the vehicle caught fire.

Recovering Dragon was not mission-critical for the demonstration flight, but Musk felt it worth trying for as he pushed toward reusability. So he authorized a small budget for the task, with the minimum team needed. The barge rental cost $40,000 a day, and a tugboat to push it to the landing site would cost as much, or more. Adding in a crew boat and other costs, every day at sea was going to run more than $100,000, exceeding what Musk hoped to spend.

"This was a lot of money for SpaceX at the time, so I was having direct conversations with Elon about it," Carlson said. "And it got to the point where we had to pay up or lose the tow boat."

Most tow boats are rated for use near land, so the closest available deep-sea ship was in San Francisco. Musk insisted on not paying for the boat until he was confident in a launch date, which kept slipping. But this put SpaceX at risk of losing the boat reservation. In early December 2010, Carlson reached a deadline he could not push off. It would take two days for the tow boat to steam down to Long Beach, and another two days to reach the landing zone. If SpaceX were going to have a barge there to retrieve Dragon, they needed to pay to play. Carlson finally got the green light from Musk at 2 AM, a mere four hours before the tugboat needed to depart San Francisco, in a terse text message: "OK," with a frowning face emoji.

The tugboat linked up with the barge in Long Beach, and the little procession began moving toward the landing zone at 5 to 10 knots early on Sunday, December 5. It felt good to finally be at sea, sailing toward the horizon and a date with history. But on that first evening the recovery team received a call on their satellite phone. Something had come up with the rocket, Carlson's crew was told. The launch had been called off.

Dejectedly, they swung the barge around and headed home.

## Emergency surgery saves Merlin

What happened? That Sunday, as a precautionary step before launch, SpaceX performed a series of video inspections of the rocket's interior. The engineer responsible for the Merlin engine that powered the second stage, Erik Palitsch, discovered a crack about three inches long running up from the base of the nozzle. Because second stage engines ignite above the Earth's atmosphere, they are optimized for performance in the vacuum of space. As a result, the upper stage engine nozzle measures nine feet long, more than twice the height of the first stage Merlin.

But even a small crack in a big nozzle could only bring bad things in flight. Most likely, as the nozzle heated up, the crack would rapidly expand, tearing the nozzle apart. The defect Palitsch found threatened to upend SpaceX's carefully laid plans to complete the critical Dragon demonstration flight before the end of 2010. Musk asked his team for options on how to proceed.

No one had any at first, but they got to work. First the engineers sought to understand the source of the problem, realizing it was caused by a one-inch line that fed nitrogen gas into the "interstage" area between the first and second stages of the rocket. The interstage, about fifteen feet tall, sits atop the propellant tanks of the first stage, and below the tanks of the second stage. NASA had been concerned about ice building up on top of the first stage propellant tank and had suggested SpaceX feed nitrogen into the interstage to prevent this. But the one-inch duct had then blown nitrogen gas almost directly onto the second stage nozzle. Over time the nozzle buckled in and out, before finally cracking.

The rocket could not fly with the nozzle as it was. So the launch team scrambled for alternatives.

The obvious solution was to replace the nozzle, and Kevin Miller called to see if NASA's Super Guppy cargo aircraft was available to transport a replacement. However, SpaceX did not actually have another vacuum nozzle ready to go, and completing the delicate fabrication process, shipping the nozzle, and installing it on the Falcon 9 would likely take a month. In SpaceX time this was an eternity, and it led the engineers to brainstorm other solutions. During the overnight hours on Sunday, in a meeting with Buzza, Shotwell, Mosdell, Thompson, and others, Tom Mueller realized that the mission might succeed if SpaceX simply trimmed several inches off the bottom of the nozzle. This would remove the cracked area and also degrade the second stage's performance. But the debut Dragon flight did not need all of the Falcon 9 engine's capability.

After everyone agreed this might work, Thompson reached for his phone. He wanted Marty Anderson, his best technician, for this job. Anderson answered the call early on Monday, December 6, in SpaceX's factory in Hawthorne as he was coming off a double shift. Thompson explained about the second stage engine nozzle and informed Anderson that he needed to fly to Florida and trim fourteen inches from the base. Anderson demurred. He loved SpaceX and would do almost anything for the company. But flying?

Prior to joining SpaceX, Anderson had worked for two decades as an airplane mechanic. A few years before the 9/11 terrorist attacks, Anderson landed his dream job as a line mechanic at American Airlines. But as the aviation industry cratered, Anderson was caught in a tidal wave of industry layoffs. Swallowing his pride, he took work on commuter buses to pay the bills. So when SpaceX came along and offered him a more glamorous line of work, he leapt at the opportunity. "I would have done that job for free, and got another job to pay my bills, just to go there," he said. "I've told Elon that a million times." Thompson came to rely on the technician's steady hands and craftsmanship, and he soon fed Anderson some of the most difficult metal work at SpaceX.

Despite his airline experience—or rather, because of it—Anderson hated to fly. "I don't hate airplanes," he explained. "But I have flown so many hours on airplanes, that I feel like my number is going to come up on the roulette wheel sooner or later."

But over the phone that Monday morning, Thompson was insistent. The nozzle issue left SpaceX hanging in the wind. Thompson needed his best guy. It was crazy as hell to be contemplating cutting off part of a rocket engine just days before flight, and SpaceX was still not sure NASA would go for it. But Anderson's skills were the only chance SpaceX had to launch Dragon that year. So Anderson relented. Thompson had saved him from a life of working on buses, so he would try and save the company a month of time. He agreed to catch a red-eye flight and arrive in Florida Tuesday morning.

A friend, another structures technician named Rick Cortez, would travel with Anderson. That day they prepared for the job, plotting the best approach to trimming the nozzle. Anderson gathered up thirty different pairs of tin snips, special tools for cutting metal, to bring with him. This got him a quizzical look when checking his bag at Los Angeles International Airport, but the attendant allowed it to be put in the plane's baggage hold.

Anderson should have been pretty mellow by this time, after forty-eight straight hours of being awake. But Cortez could plainly see his friend's anxiety about boarding the flight. He persuaded Anderson to stop at one of the airport bars, where they pounded down a couple of beers before takeoff.

This had the desired effect, and the flight went smoothly.

After renting a car in Orlando, Cortez drove the pair to Cape Canaveral. Stopping at SpaceX's control center, Anderson briefly met with Thompson, Buzza, and other company officials. He had three questions: Was the rocket fueled? Had its flight termination system been activated? Was the booster powered up? Satisfied with negative responses to his queries, Anderson donned Carhartt overalls and proceeded to a crane. There he met Florence Li, one of Thompson's key structures lieutenants. They rose 100 feet into the air, toward a hatch leading inside the interstage.

Riding in the crane bucket, Anderson noticed the cold. The temperate Central Florida region had recorded a rare freeze that morning, and a blustery northerly wind still blew across the Space Coast. Anderson felt relief as he bent down to enter the rocket and get out of the wind. Inside, he took a moment to marvel at the reflective interior walls and domed ceiling. The Merlin engine with its expanded nozzle hung above him, from the top of the interstage, as if it were a church bell. "It looked like a cathedral inside," Anderson said.

Then Anderson got to work. He started by tracing a line around the nozzle's circumference, fourteen inches above its base. He then took a larger pair of tin snips and started to cut an inch or so below this line. Like most spaceflight hardware, the second stage Merlin engine nozzle was optimized for the highest possible performance at the lowest weight. To fabricate the second stage nozzle, SpaceX had selected a metal alloy composed mostly of niobium, which is as hard as titanium and can withstand temperatures that boil steel. Although the nozzle's material was only twice the thickness of a soda can's, the metal was still very rigid. This meant that as Anderson started to trim the nozzle just below his line, he could not simply pull away the pieces he cut. He had to snip eight or ten inches, and then cut downward to remove pieces.

As he snipped away parts of the nozzle, Li would hand them outside to a pair of engineers waiting in the crane's bucket. Anderson felt bad for them out in the cold weather, while he and Li were in comparative comfort. But he was suffering, too. The nozzle had a diameter of eight feet at its base, so Anderson had to cut a total of twenty-five feet around the engine. Halfway into the job, Anderson noted some pain in his cutting hand. He looked down and saw that the handle of his tin snips had worn through the glove he wore, as well as his skin. He could see a dime-sized area of white bone. Li urged him to take a break, but Anderson asked her to pass the duct tape, which he wrapped around the injury.

Anderson finished about four hours later, having made his initial cut, and then gone around and trimmed it with his smoothest pair of tin snips. This ensured a nice, straight line around the shortened base of the Merlin engine. Before exiting the interstage, Anderson stopped to admire his handiwork. Looking at the smooth edge, he could not even tell it had been cut. Even so, SpaceX still had plenty of work to do before the second Falcon 9 could leave the launch pad. Although Anderson may have been satisfied with the resized nozzle, NASA still needed convincing.

After partnering with NASA since 2006, SpaceX had built up a reservoir of trust with the space agency officials who worked most closely with the company. A NASA engineer delegated to manage the company's contracts with the space agency, Mike Horkachuck, had been embedded with SpaceX engineers for four years, and he understood the company's technology inside and out. And Alan Lindenmoyer, NASA's manager of the commercial cargo program, had also come to believe that SpaceX was on the right track. In advance of the Dragon launch, Lindenmoyer arrived in Florida on Monday morning for a Launch

Readiness Review meeting. During this meeting, in which senior officials discuss any open issues prior to a launch, SpaceX engineers disclosed the problem with the nozzle. They played the video showing the purge vent blowing nitrogen onto the nozzle, which fluttered like a paper in the wind.

"They had already figured out the problem," Lindenmoyer said. "The question was, what were they going to do about it? We assumed they were going to have to roll the vehicle back to the hangar and replace the nozzle. And this would be weeks and weeks of delays."

As SpaceX engineers explained their proposed solution to snip the nozzle, Lindenmoyer's initial reaction was shock and disbelief. Coming from the culture of NASA, he knew the agency would never take such a risk. Its engineers would have completed a full analysis over a period of weeks, then gone through an extensive testing and certification process. These were lessons hard won by the space agency after decades of spaceflight and the loss of three crews dating back to the Apollo 1 mission.

And yet Lindenmoyer's charge was to be open-minded toward NASA's commercial space partners. The Bush administration was determined to give U.S. industry a shot at delivering cargo to the space station, with a new way of doing business. Instead of telling industry what to build, NASA specified what service it wanted to buy. This gave industry more flexibility to do things their way.

To make this point, when Lindenmoyer made his quarterly visits to SpaceX, he made sure not to sit at the head of the table. "I made it very clear to Elon and his team at that first meeting in 2006," Lindenmoyer said, "they were in charge. We were an investor, and all we wanted was to see them succeed."

On its first flight, Cargo Dragon was not going to the space station. Rather the goal was simply to put the spacecraft in orbit, test out its handling capabilities, and bring it back to Earth. Because this Dragon was not flying to its station, it meant that technically NASA could not actually tell SpaceX what to do about the cracked nozzle.

After the briefing in Florida with SpaceX, Lindenmoyer remained in Florida while Horkachuck flew to Washington, D.C., to brief the agency's senior leadership the next day. There were two key officials present. Bill Gerstenmaier, who managed human spaceflight operations for NASA, was there. However, since this mission would not visit the space station, it did not fall directly into his portfolio. Rather, the presiding NASA official was an old-school engineer named Doug Cooke, who oversaw exploration systems.

Cooke cut his teeth during the space shuttle's development in the 1970s. He worked in the shuttle program office during the space shuttle *Challenger* disaster in 1986 and was a technical advisor to investigators of the *Columbia* accident in 2003. Three decades of flying the shuttle had made him risk averse.

Because of SpaceX's freedom to design Dragon to its own needs, senior officials like Cooke were not particularly well-versed in the company's technology. But Cooke trusted Horkachuck, who reviewed the technical information SpaceX had provided regarding the nozzle surgery. Yes, trimming several inches off the bottom of the nozzle would reduce the Merlin engine's capability, Horkachuck said. But because this Dragon spacecraft was not traveling to the station, it did not need all of that second-stage performance. Given the high thrust chamber pressure, there were also concerns about the nozzle's structural integrity. But at the end of his presentation, Horkachuck said he had confidence in SpaceX's proposed solution.

Cooke did not share this confidence and was not comfortable with the company's decision. "I'd been through a lot of stuff on the space shuttle," he said. "And this just seemed a little too out of the ordinary." He was in favor of SpaceX taking the time to replace the nozzle and said so. But Cooke also recognized the limits of NASA's power to compel SpaceX to do anything. "This was not our vehicle," he said. "So it was not really our choice." Cooke told Horkachuck that the decision, and the consequences of that decision, should be left to SpaceX.

Later on Tuesday, back in Florida, avionics VP Hans Koenigsmann held a final roundtable meeting to assess readiness for the launch. One by one, he went around the table, asking people like Buzza, Mueller, and Thompson if the vehicle was ready to fly. They all confirmed that their departments were "go" for launch. Then, Koenigsmann turned to Lindenmoyer, who saw himself as just an observer in the room.

"It surprised me when Hans turned to me," Lindenmoyer said. "I made it clear NASA did not have the authority. But what I did say is we did not object to their plan."

Lindenmoyer understood the company's desire to launch. SpaceX had hired madly after winning the two NASA contracts, growing to more than 1,000 employees as it simultaneously built a large new rocket and spacecraft. However, the company would receive the bulk of its funding from NASA only after reaching certain milestones. For this first demonstration launch, SpaceX would get paid as long as the rocket cleared the launch tower in Florida. A delay of weeks, or months, would push off a much-needed infusion of cash to keep the company's projects rolling forward.

Lindenmoyer also understood there would be consequences for a failure. He ran a lean office, with fewer than ten people to manage contracts with SpaceX and Orbital Sciences. With his program on NASA's back burner, Lindenmoyer and SpaceX could fly under the radar. This limited interference from skeptics of SpaceX and commercial space, both at NASA and in Congress. There, lobbyists for the traditional space contractors whispered that Musk could not be trusted. He was reckless, they said. By cowboying a fix to the nozzle, and then failing, SpaceX might shoot itself in the foot.

"I was trying to keep NASA's involvement with SpaceX at arm's length to let them innovate," Lindenmoyer said. "But a failure would have really thrown a wrench into it. I do think we, and they, would have been criticized. Congress would have wanted us to take a much more hands-on approach."

As day turned to night on Tuesday, with the launch scheduled for the next morning, these thoughts swirled inside Musk's mind. While weighing the pros and cons of lifting off with the snipped nozzle, Musk reached for his mobile phone to call the engineer at NASA he trusted most: Bill Gerstenmaier. He was nearly a generation older than Musk, having started his career at NASA in 1977, when Musk was only six years old. But by late 2010, both men respected one another. Even though Cooke had oversight of the launch, Musk wanted Gerstenmaier's endorsement of the plan. Was this plan stupid, Musk wanted to know?

Gerstenmaier had looked at SpaceX's data, and for him data mattered above all else. SpaceX's modifications to the upper stage should pose no issues on the next day's launch, he said. This reassurance helped Musk feel more confident. After hanging up with Gerstenmaier, Musk telephoned Buzza. "We are go for launch," Musk told his launch director.

## Snagging the Dragon from the sea

Hours after Carlson and the recovery crew received the bad news about the damaged nozzle and the need to turn around at sea, they received another call on their satellite phone: sit tight and hold your position. Early the next morning another call came. Somehow the nozzle issue was going to be solved. They were to push onward toward the recovery area.

"We were all just like, you know, fucking SpaceX," Carlson said. "They're just out there doing one more incredible thing. We didn't know how they could do this. We didn't know how they could fix something that seemed impossible."

Due to the delay, the recovery fleet would stay at sea for a few days longer than planned. The tugboat had plenty of diesel fuel, but the boat carrying the SpaceX team, the eighty-five-foot-long *Gladys S* workboat, was not provisioned for an extended cruise. So the captain agreed to tie the crew boat to the barge, thereby letting the tug pull both vessels. With its engines shut off, the *Gladys S* went eerily quiet that night as the dozen SpaceX employees bedded down on the lower deck. On small, primitive beds that had been hastily framed from two-by-fours, they fell asleep on the rolling seas. Then, in the middle of the night, a tremendous bang shattered the silence. The *Gladys S* started rolling. The small galley where the crew made meals tumbled over, and the boat's crew scrambled onto the deck. Amidst the waves, the clevis fastener at the front of the crew boat had broken off, severing the link to the tugboat. Soon the *Gladys S* engines were powered back up, and everyone settled back into an uneasy sleep.

This hardly proved an ill portent, however, as the ocean turned calm on Tuesday morning when the recovery team arrived at the landing zone. The SpaceX crew transferred over to the barge, making final preparations, such as practicing with the crane and two Zodiac inflatable boats.

Carlson and his team had arranged all sorts of backup plans for contingencies, including a runaway Dragon. SpaceX had performed landing tests near Morro Bay, about halfway between Los Angeles and San Francisco. During one of these tests, the parachutes, on a particularly windy day, had remained inflated after a Dragon capsule was dropped by a helicopter from 14,000 feet into the bay. Buoyed by the breeze, Dragon started kitesurfing across the ocean, heading almost directly toward Diablo Canyon Power Plant, the last nuclear power facility in

California. A support helicopter photographing the test had to fly near the surface and disrupt the parachutes.

Uh-oh: During a parachute test, the wind drags Dragon towards the Diablo Canyon nuclear power plant. | PHOTO CREDIT: ROGER CARLSON

Leading Dragon recovery meant that Carlson spent nights worrying about such things happening when Dragon returned from space. What if the parachute problem happened at sea, and Dragon slammed into the barge or tugboat? They would have no helicopters to beat down the chutes. His solution called for a far simpler flying object—a T-shirt. Carlson and the recovery team purchased air cannons, the powerful ones used to shoot souvenir T-shirts into the upper deck at sporting events. They brought heavy wads of material, intending to shoot them into any inflated parachute, loading the chute up so it would deflate and collapse. They even practiced shooting T-shirts, immensely enjoying the knowledge that they would expense the activity.

After rehearsing on the day before Dragon's flight, Carlson and a few others gathered in the wheelhouse of the tugboat to make their final plans for the next morning. As the last bit of the sun dropped to the horizon, they saw a rare optical phenomenon known as a green flash. "For two seconds, the whole world turned green," Carlson said. "I thought it was a myth. But the ocean was so flat that it was like a searchlight coming on. I took that as a good omen." Green meant go.

After the SpaceX recovery team awoke the next morning, they got the word from Hawthorne via satellite: the second flight of the Falcon 9 rocket, bearing the Dragon spacecraft, had launched. About seventy-five minutes after Dragon soared into orbit on the Falcon 9 rocket's second stage, a pair of engineers on the *Gladys S*, named Eric Hultgren and Eric Massey, picked up a telemetry signal as the spacecraft flew overhead. This signaled that all was well, and that after one more spin around the planet Dragon would descend toward their small patch of ocean. The skies were fair as Carlson deployed the two small Zodiac boats to the western edge of the recovery ellipse, a dozen miles away, with the barge stationed on the east side.

Kevin Mock admired the turquoise water as his Zodiac zipped across the slowly rolling ocean. The swells were ten feet tall, but during the thirty-second periods between waves the surface was like glass. The Zodiacs were soon far enough to the west they could not see the barge or any other boats on the horizon. Mock's only connection to the world was a radio, calling out Dragon's descent milestones. He had joined SpaceX less than a year earlier, after earning an aeronautical engineering degree from Embry-Riddle Aeronautical University in Florida. Mock moved to Texas in January 2009, helping to test the thruster system that propelled Dragon in space. In comparison to the Merlin rocket engine's roar, with more than 100,000 pounds of thrust, Dragon's propulsion system whispered. Each of the spacecraft's sixteen Draco thrusters had just ninety pounds of thrust, or less than 1 horsepower. But in the microgravity of space, even this small amount of thrust provided plenty of oomph for maneuvering.

Now Mock led his small propulsion team into the middle of the ocean to "safe" Dragon after it landed. This entailed securing any leaking propellant and emptying its fuel tanks. This was a dangerous job. Like most spacecraft and satellites, Dragon's thrusters used hypergolic propellants, which spontaneously combust when mixed. Turning on a thruster, therefore, was as simple as opening and closing the propellant valves inside. The fuels used by Draco had another big advantage, too. The monomethyl hydrazine fuel, and its nitrogen tetroxide oxidizer, were storable for long periods of time at room temperature. This made them ideal for spacecraft.

There was a catch, though. There is always a catch in spaceflight. Hypergols are extremely toxic. In 1960, at the Soviet Union's Baikonur Cosmodrome, an accident occurred while workers were handling the hypergolic-fueled second stage of an R-16 intercontinental ballistic missile. The second stage engine ignited as the official in charge of the Soviet strategic rocket forces, Mitrofan Nedelin, urged his personnel to meet a launch deadline. The incident killed as many as 300 people, making it the deadliest rocketry accident in history, known today as the Nedelin catastrophe. Mock's job was to prevent such an accident with Dragon in the face of some key unknowns. First, he did not know how well Dragon's plumbing and fuel tanks would weather the fiery reentry through Earth's atmosphere. Additionally, he and other SpaceX engineers were not sure how much fuel would remain onboard. They estimated the volume at more than 100 gallons of hypergols.

Their first indication of Dragon's imminent arrival came from two sonic booms in quick succession, as the spacecraft slowed beneath the speed of sound. Then Mock and another engineer, as well as two hired divers, saw the spacecraft's small drogue parachutes appear, followed by deployment of the three mains. Dragon splashed down about a mile away from them, and the Zodiacs revved their engines to chase down the spacecraft. They noticed, to their relief, that Dragon was not sinking. That was a good sign for its structural integrity. Mock then deployed a twenty-foot painter's pole with a hypergol sniffer on its end. Mock radioed back to the barge and crew boat that all looked good, and there were no leaks detected. After deploying some flotation rafts around the spacecraft, they began towing it toward the barge.

As Dragon neared, Carlson watched with equal measures of relief and concern. Although Dragon floated, he still worried about it being too waterlogged for the improvised crane to lift onto the barge. One particular scene from a James Cameron movie, *The Abyss*, kept running through his mind. In the science fiction film about an unexpected encounter in the depths of the ocean, a violent storm engulfs the exploration vessel. Amid the carnage, the entire crane and cable system break off the boat and fall into the water. "I had nightmares about that crane falling into the ocean, and dragging Dragon down with it," Carlson said.

He need not have worried. None of the contingency plans were needed as the crane pulled Dragon from the ocean by its docking ring mechanism and set it into the nest. By that point it was 1 PM local time, and the seas had started to come up. Sunset would arrive in less than four hours, and there was just enough time to bolt down Dragon and for Mock's team to vent the pressure from the propulsion system. They did this work in protective suits, with independent air supplies, as if they were dealing with a biohazard during a pandemic. With this done, the spacecraft was in a safer condition. But two days of work remained to offload all of the propellant.

Before they left the barge for the crew boat that night, the recovery team took a few minutes to marvel at Dragon. Only hours before, it had been in Florida, a gleaming, shiny white spacecraft on top of a rocket. Some of their friends had been standing next to it. Now it had traveled around the planet once and nearly completed a second orbit. It looked like a slightly charred marshmallow. Some of the paint had flaked off. But it was in one piece. And it smelled fantastic, having that toast-like odor that's characteristic of metal having flown through the atmosphere. They took in the scene, then left to crash overnight in their small bunks on the *Gladys S*.

The next morning the propulsion team transferred from the crew boat back to the barge, beginning their work to offload the hypergolic propellants. Getting onto the barge while at sea was an ordeal. It involved the *Gladys S* essentially ramming into the side of the barge, where there were large truck tires fastened to absorb the collision. Then, one by one, personnel would carefully clamber up a rope ladder onto the barge. (During a later recovery mission, a SpaceX engineer lost a foot hopping from boat to boat.) Once transferred over, the six engineers and technicians led by Mock worked for a good eight hours offloading the oxidizer from each quadrant of Dragon, filling special hazard containers. After this flight, SpaceX would receive approval from the U.S. Department of Transportation to move Dragon by road with these fuels still onboard, provided the propellant system was depressurized. But this first time, the dirty work would be done at sea to avoid putting the public at risk.

As Mock's team worked, the weather worsened. The barge began plowing through big waves, sending a shower of salt spray running along its length. Their inflatable suits kept the water from the workers, but it was increasingly cold and slippery on the barge, and the night turned dark before they were finished. The captain of the *Gladys S* approached the barge to pick up the recovery team but determined that sea conditions were too dangerous for a transfer.

This was one contingency Carlson had not planned for, because the barge was not rated for overnight stays. So he improvised. The SpaceXers back on the crew boat gathered up nonperishable snacks, such as granola bars, and a thermos of coffee into a trash bag. Then they rolled up six sleeping bags in another. As the *Gladys S* pulled near the barge, the bags were heaved over. The recovery team—Mock and another engineer named Michael Altenhofen, and three former Marines who were technicians, Don Bell, Jacob Foster, and Walter Gonzalez—retreated into the twenty-foot shipping container that had been outfitted as a rudimentary Dragon control center. Bone-tired from their exertions all day in pressure suits, they ate their snacks for dinner and zonked out.

"We got the best night's sleep we could, with six dudes sleeping next to one another on a wooden floor," Mock said. "We were exhausted, so I can't say that I slept terribly that night."

The next morning Mock and his team resumed their offloading operations. Watching these exertions from the *Gladys S*, Carlson was amazed. Some of his team members on the crew boat were retching and weak from sea sickness. But Mock's engineers kept powering through their difficult and intense work. "It's the hardest work I've ever seen at SpaceX, or anywhere," Carlson said. The following day the small fleet pulled into the Port of Long Beach, their prize intact. Hordes of their colleagues had made the twenty-mile drive from SpaceX's factory to greet the recovery team as returning heroes, whooping and hollering and sharing the moment of Dragon's triumphant return home.

Less than two weeks later, cleaned up inside and out, Dragon played a starring role at the SpaceX Christmas party. The company placed giant curtains around the vehicle, with one glass wall open. At the party, revelers could walk along a corridor to the back of the factory to see Dragon highlighted by multicolored stage lights.

Dragon deserved its place of honor. In more than half a century of spaceflight no private company had ever built a spacecraft, launched it into orbit, and recovered it as SpaceX had just done. Mock lingered for a long time near the showroom. In college, he had interned at NASA's Kennedy Space Center. It seemed like a dream assignment, getting to work on the Space Shuttle Program, and seeing all that incredible space hardware up close. But pretty quickly, he realized NASA was not for him. As an intern, he had not felt as though he had done anything meaningful. Now, in just his first year at SpaceX, he had gotten his hands dirty and made a little history. "I sat there, by myself, staring at this thing that we had just pulled from the ocean," Mock said. "I was in awe of what had happened."

For Carlson, too, Dragon's recovery capped one of the best periods of his life. He had been born not long after the Apollo 1 fire in 1967, a horrific tragedy that killed three NASA astronauts trapped inside the spacecraft. His parents named him after Roger Chaffee, the mission's young pilot. Carlson grew up watching the Saturn V launches that sent astronauts to the Moon. As much as he admired their courage, what had struck Carlson most were the landings. He loved seeing the Navy frogmen leap from helicopters to help astronauts out of the space capsule. "I knew that's what I wanted to do when I grew up," he said. Carlson had not dived in the water to recover Dragon, but he'd led the operation. For the holiday celebration he wore a bright red suit, a bowtie, and an ear-to-ear grin. It was the best Christmas party ever.

SpaceX eventually hung the prized spacecraft just outside the Dragon Mission Control Center in Hawthorne. During crew launch broadcasts by NASA and SpaceX, it can be seen in the background, behind the hosts. More than a dramatic showpiece, Dragon remains a silent sentinel to this day, ever watchful, as if ensuring that the rest of the brood follow it safely home.

|5|

# FLIGHT THREE: DRAGON'S DESPERATE RIDE

*May 25, 2012*
Houston, Texas

It was nearly there.

In the blackness of space, the shiny white Dragon spacecraft hovered tantalizingly close to the International Space Station, a mere 250 feet away. But it could not approach any nearer without permission from NASA. And with the lives of astronauts onboard at risk, the space agency could shut Dragon down at any moment.

This seemed imminent, and tensions were running high in Houston and Hawthorne.

Edgy and anxious, Musk and the SpaceX engineers responsible for flying Dragon watched their spacecraft from Hawthorne. They agonized as seconds ticked away and Dragon drained precious fuel to hold its position. Ten minutes passed. Then twenty. Dragon remained in a holding pattern because its sensors were having difficulty locking onto the space station. As a result, the spacecraft could not precisely determine its position.

As Dragon's date with destiny slipped away, its fate lay in the hands of a thirty-eight-year-old mechanical engineer in Houston. Inside NASA's Mission Control, flight director Holly Ridings faced a difficult decision. She could break the mission's flight rules and allow Dragon to advance toward the station despite problematic sensors. Or she could make the safe choice, cutting Dragon loose and ending its flight. Many of her senior colleagues would not have wavered, taking the least risky option for the three astronauts onboard the station. But she wanted to give SpaceX more time.

Typically, when directing a flight, Ridings paced at the back of Mission Control. But on this morning in late May, she sat at her desk, focused on the data before her, with communication loops buzzing through her earpiece. "I don't like to sit down," she said. "Normally I just pace, pace, pace. But I was sitting, with absolute total focus. All of my concentration was balled up into that moment, just razor sharp."

Dragon had launched three days earlier, on the third flight of the Falcon 9. It soared into orbit, flying much higher and farther than Dragon's first test a year and a half earlier. Now came the most difficult part of the mission, an attempt to berth Dragon to the station. Ridings and her team of flight controllers had arrived on console well before midnight for the operation. By sunrise the next day, efforts to capture Dragon with the space station's robotic arm were running hours behind schedule.

Ridings had only minutes to decide whether to let Dragon proceed, but she had trained for this moment her whole life. She grew up in Amarillo, a small city in the Texas Panhandle made famous by country musician George Strait. As a kid in the sixth grade, she and her classmates had filed into the cafeteria to watch a historic space shuttle launch. A teacher rolled one of those big, boxy televisions in on a cart, and the good children of Amarillo watched as space shuttle *Challenger* broke apart just seventy-three seconds into its flight, killing schoolteacher Christa McAuliffe and six other astronauts. The shocked silence in the cafeteria, followed by tears on the faces of the teachers, left a strong impression on Ridings. In her twelve-year-old mind she decided to do something so a horrible thing like that never happened again.

Flight Director Holly Ridings reviews data at her console in 2010. | PHOTO CREDIT: NASA

A little more than a decade later she made good on that childhood promise, becoming a civil servant at NASA on the brand-new International Space Station program. Soon she took a step up, becoming a flight director. After leading a space shuttle mission to the station in 2009, Ridings considered options for her next assignment. Her husband, a fellow NASA engineer named Michael Baine, suggested she work on a commercial mission. He thought that was the next big thing.

For a time, Ridings worked with both SpaceX and Orbital Sciences, which was also developing a cargo spacecraft for NASA. In 2011, when SpaceX got closer to its first flight, she transitioned to become flight director for Dragon's first flight to the station. A quarter of a century after *Challenger*, her job was quite literally to make sure something horrible didn't happen, such as Dragon smashing into the station.

This all floated in the back of her mind as Ridings considered SpaceX's solution for Dragon's dysfunctional sensors. It involved rewriting the vehicle's flight software on the fly, and that's just not something that NASA did. "In our world, doing anything to the software that close to the space station gets our antennas up," Ridings said. "It is pretty concerning."

As the minutes melted away, Ridings turned around and consulted with Norm Knight, the chief flight director. She knew that NASA's manager of the International Space Station, Mike Suffredini, was following along closely with the agency's other senior leaders down the hall. But in that moment, in that room, Ridings had the final say. Her word might as well have been the word of God.

As is standard before important events in spaceflight, Ridings polled the flight controllers on her team to see if they were go or no-go. Some members of her team in the control center were no-go. Therefore, if Ridings allowed Dragon to fly onward, she would have to go against people who worked for her. That kind of thing rarely happened in Mission Control.

As Dragon held its position, Ridings kept a line of communication open to her counterpart in Hawthorne, a former U.S. Navy pilot named John Couluris. He had come to SpaceX five years earlier to help get Dragon to the space station, and then bring it safely home. For three of those years, Couluris had worked closely with Ridings, building up the level of trust needed now. Quietly and calmly, he told Ridings over the private communications loop that SpaceX had this.

But did they? For Ridings, a wrong decision would not break her career. But if things went sideways, it would set her back; and not just her, but potentially other women at NASA as well. Ridings was not the first female flight director at NASA, but of the five dozen people who reached that revered position before her, only five had been female. She carried the burden of other women trying to prove themselves in the male-dominated space industry.

With time running out for SpaceX and Dragon, Ridings took a deep breath and made her call.

## "The look of horror was very, very real."

Couluris hailed from New York, the son of a long-time engineer at Northrop Grumman who worked on the lunar module that carried Apollo astronauts to the Moon. His father's experiences left a deep imprint, including a love of spaceflight and a desire to serve his country. After earning a graduate degree in aerospace engineering at Rensselaer Polytechnic Institute, Couluris joined the U.S. Navy and began flying the P-3 surveillance aircraft.

During the 1990s, he piloted patrol missions for antidrug operations in South America and tracked enemy submarines from a base in Iceland. At the end of the decade he commanded flights carrying cruise missiles out of Sicily that overflew Kosovo as part of the American effort to stall the invasion by Yugoslavia. After earning two Air Medals, Couluris became a test pilot for the Navy. Then some of his squadron mates started leaving to fly for a new airline, JetBlue, and he decided to join them. Based on Long Island, a JetBlue job allowed Couluris to return to near where he grew up.

JetBlue was a scrappy upstart in the early 2000s, and when Couluris signed on he was among its first few hundred pilots. The startup faced stiff headwinds when larger airlines, like Delta and United, created low-cost regional airlines to crush the new venture. Using his programming skills, Couluris wrote some code to scrape the seat maps put online by JetBlue's competitors to see how full they were and what prices they charged. This allowed JetBlue to price its seats accordingly. It also earned Couluris a promotion to management, where he eventually led crew operations.

Couluris joined JetBlue, in part, to collect stock options with the intent of cashing them in one day to start his own space company. However, in 2006, when the price of oil hit $100, these options were worthless, and his dreams of becoming a space entrepreneur were shattered. Still, he longed to get into his father's industry. Couluris had been tracking the exploits of SpaceX, so at the end of the year he sent in a résumé.

Musk called a few weeks later. He liked that Couluris came from JetBlue, a leader in price and on-time performance. Musk wanted SpaceX to emulate the airline industry's operations, with frequent flights and minimal turnaround between missions. He asked Couluris to oversee recovery operations for rockets and spacecraft, while also standing up a mission operations department. When Couluris joined SpaceX in April 2007 he was thirty-six years old, about a decade older than the company average.

One of his first jobs entailed convincing NASA that SpaceX could safely operate Dragon in orbit. During his initial trip to the Johnson Space Center in Houston, Couluris spoke to the Mission Operations Directorate, which bore responsibility for training astronauts and flying them safely into space and back. He tried to explain SpaceX's then-unorthodox philosophy. Musk did not want to build a spacecraft and sell it outright to NASA. Rather, he wanted to build the spacecraft and charge NASA a fee to fly its cargo. It's like Fedex, Couluris explained. You provide us a package, and we'll deliver it to space for you.

"This seems obvious today, but the look of horror on their faces was very, very real," Couluris said. "They were like these guys, and that guy from PayPal, are going to approach our $100 billion station with six astronauts onboard?"

As it had done with the Falcon 9 and its launch pad, SpaceX looked to scrappily build as much of Dragon in-house as possible. For example, to keep science experiments cold in refrigerators during the ride into space, Dragon needed powered lockers. A locker required two latches to open and close each compartment, and NASA's supplier wanted $1,500 per latch. This cost posed a problem until a SpaceX engineer found inspiration while contemplating the latch on a bathroom stall. Perhaps, he wondered, Dragon might be able to make do with a similar latch? So with $30 in parts, the company fabricated its own locking mechanisms that proved more reliable than the expensive, aerospace-rated version.

Additionally, to strap down large bags in Dragon's cargo racks, SpaceX first tried the space-rated straps NASA used for similar compartments in the space shuttle. They cost a fortune, so the engineers searched for alternatives, eventually finding a solution in NASCAR seat belts. SpaceX got paid when it completed milestones, such as proving the viability of its makeshift solutions, and these payments were critical to the company's viability. So Couluris's operations team halted their training to fly Dragon in space to help the structures team build and test the first cargo racks. Although they lacked the relevant training, the operations engineers riveted and welded and sanded together an iron structure to meet NASA's requirements and unlock a payment.

"In almost everything they were designing and building cost was a significant factor in their choices," said Mike Horkachuck, the NASA official designated to manage SpaceX's cargo contract. The SpaceX services model was unlike a traditional cost-plus contract, in which NASA reimbursed a contractor for costs, plus a fee.

"Under a cost-plus contract, a lot of the engineers didn't talk so much about how much it would cost to build a part they designed," Horkachuck added. "But SpaceX was always considering whether to make a part or buy it, and when they would get the cost in for an aerospace quality part, they would say no way am I paying $50,000 or $100,000 when I could build it in my machine shop for less than $5,000."

As part of the certification process, SpaceX had to convince NASA that its scrappy substitutions were up to the agency's rigid safety standards. Dragon's first demonstration flight in 2010 faced far less scrutiny as it was not intended to fly anywhere near the station. Because of this, Doug Cooke had been powerless to stop the launch with a snipped Merlin nozzle. But no longer. Dragon's second flight would be going to NASA's most valuable asset. To do this Dragon would need to be a fully mature spacecraft, with ample redundancy in navigation and communications systems, and the ability to withstand multiple failures. At any point, a litany of NASA officials could shut the mission down.

The first person in line to make that call was Kathy Lueders.

## Do this for Kathy

Lueders attended New Mexico State University in the 1990s, studying industrial engineering, before starting her NASA career nearby at the White Sands Test Facility. A few years later she moved to Houston and worked on the International Space Station. In 2006 the program manager, Mike Suffredini, asked Lueders to draft "requirements" for private vehicles visiting the station.

Requirements are tedious, and for a time they were the bane of SpaceX's existence. But they are the bureaucratic currency by which

NASA buys off on the safety of a spacecraft. A requirement could be anything from determining where a spacecraft must launch from, to defining the amount of stowage it should have, to ensuring there are no sharp objects inside a vehicle.

NASA wanted SpaceX (as well as Orbital Sciences and other commercial spaceflight companies) to succeed, so Suffredini urged Lueders to make the requirements on Dragon "lean." She took this to heart and pushed back on mid-level managers at the space agency seeking to levy additional rules and requirements on the private companies. Mostly, she succeeded. Whereas the space shuttle had more than 10,000 requirements, Dragon ended up with about 400.

During an early meeting with SpaceX when its offices were still in El Segundo, Lueders invited flight directors and others from NASA's Mission Operations to join her. The NASA people, she thought, could share their experiences. Many of the government participants brought along folders, and at the end of the day more than two dozen binders were stacked on the table, each providing insight into NASA's best spaceflight safety practices. When the flight operations officials left and Lueders remained alone with the SpaceX team, she knocked the tower of binders down.

"Here's the deal," she told them. "You guys have to go figure out how to do it your way."

Lueders developed a close relationship with Gwynne Shotwell, who regularly attended the Dragon meetings. The two women, along with NASA's Alan Lindenmoyer, held frank and open conversations about the problems SpaceX was having. These issues were both technical and financial. An engineer with tremendous social skills, Shotwell spun a lot of plates for SpaceX by leading sales, managing multiple directors, and trying to keep the company's books in order. She had come to SpaceX near its beginning, to help Musk lead the spaceflight revolution. They made a dynamic pair. Musk said crazy shit sometimes, but Shotwell's word was trusted by NASA and other customers. Whereas Musk could be abrasive, she was the velvet glove so important to wooing and retaining customers. Yet anyone pushing her too far would find steel underneath that winning smile.

By winning the cargo development contract from NASA in 2006, and later the money for operational flights, Shotwell not only financially saved SpaceX but set a foundation for all that would come. "These programs were defining for SpaceX's success," she said in an interview. "We would not be the company that we are today without having won them and done that work in partnership. That is a true partnership with NASA."

This partnership extended to sharing the extent of SpaceX's financial struggles with Lueders and Lindenmoyer. During quarterly meetings, she and Lindenmoyer would retreat into a private conference room to candidly discuss SpaceX's burn rate and cash position.

"They were really walking a tightrope," Lindenmoyer said. "Gwynne had a spreadsheet she managed personally on her computer, where she tracked the financial status of the company. She would share exactly what she was seeing. And I have to say, sometimes I would leave these meetings thinking, *Oh my goodness, they're going to run out of money.* They were depending on revenue sources that weren't yet inked. Gwynne was making deals, but the company was growing rapidly and the burn rate was increasing."

Shotwell and Lueders also refereed disputes between technical teams at SpaceX and NASA. Shotwell managed the Dragon team, led initially by Steve Davis and, later, David Giger. As Giger, Couluris, and others argued for leniency on a particular requirement, Shotwell would make SpaceX's case to Lueders if she felt it was just. Back at NASA, Lueders had her own disagreements with "requirement owners" at the agency. She earned credibility with SpaceX by asking tough questions of her own people. Shotwell came to trust her NASA counterpart and would tell the Dragon team that they had to "do this for Kathy," knowing how hard Lueders was fighting the traditional space culture in Houston.

"Gwynne was in the middle," said Bulent Altan, vice president of avionics. "We were pushing from one side, and Kathy the other side. We were at odds, but that really resulted in a good capsule."

In the spring of 2010, SpaceX finally hired someone to deal directly with requirements. Abhi Tripathi had been a civil servant at NASA, working on the Constellation Program intended to return humans to the Moon. Several hundred people worked at NASA to make sure Constellation's rockets, spacecraft, and landers met the agency's requirements. But at SpaceX no one had been tasked specifically to understand and coordinate requirements for Dragon. This part of aerospace engineering fell into the category of "systems engineering," a job title explicitly disallowed at SpaceX by Musk.

"During my orientation I discovered someone else was there who had been hired to do the same job as me," Tripathi said. "We were both hired the same day, for the same job. Do you want to know why? It's because they figured that one of us would quit, or be fired, after about a month."

He soon understood why. After his orientation, Tripathi took a tour of the factory, viewing the Dragon spacecraft in a clean room. Then he received a spreadsheet that was the company's requirements tool and was told to demonstrate to NASA that the Dragon spacecraft met all of the space agency's requirements.

"At NASA nothing got built without thousands of meetings for each requirement," Tripathi said. "You don't take action until everyone signs off on the requirement, and changing a requirement would take months. And here I am looking at a mostly built spacecraft."

Tripathi did not quit. Instead he dug in, meeting with the people building various parts of the spacecraft. About a third of the engineers designing and building Dragon knew about the requirements and had done their best to meet them. But two-thirds of the engineers had not and simply worked on what they believed to be the best design. So when Dragon engineers saw Tripathi coming down the line, they were not well pleased. It meant more work.

NASA was equally wary. The agency's engineers were accustomed to receiving hundreds of pages of documentation from contractors to verify a spacecraft met a particular requirement. At SpaceX, an engineer might provide Tripathi with a screenshot of the design for a part, and perhaps half a page of text.

"At the beginning I don't think we had a lot of respect for what NASA wanted, and what we were doing did not inspire a lot of confidence from NASA," he said. "Eventually we ran out of goodwill. I had to vouch to Kathy that we would not submit 'crap' any more. I literally used that word."

Even so, NASA officials working with the company came to realize its engineers were serious about spaceflight. To demonstrate that Dragon met a requirement for low ambient noise in the vehicle, for example, SpaceX told NASA it had to conduct the test at 2 AM on a Sunday morning. This was the only time when the big welding machines were turned off and the factory was quiet enough.

The SpaceX engineering teams had little time for paperwork because they had so much hardware to work on. The first Dragon mission, known internally as C1, included a lot of shortcuts because it only remained in space for about three hours. For C2, though, Dragon had to be completely remade into a sophisticated vehicle that could fly all the way to the International Space Station. This involved adding triple redundancy to its avionics, installing cargo racks, developing a new trunk with improved solar arrays, adding an active temperature control system, and much more.

Version 1.0 of the Falcon 9 rocket could only launch Dragons, so all of the schedule urgency fell onto the shoulders of the Dragon team. There would be no more launches until C2 was ready to fly. Eventually the gap between launches stretched to eighteen months, SpaceX's longest ever since the Falcon 1 rocket's debut six years earlier. For Giger, this year-and-a-half break between C1 and C2 was the most difficult period of his life.

"There was immense pressure on the Dragon team to move faster," Giger said. "We completely redesigned the spacecraft, and this was a monumental undertaking. I think I worked more than 320 days during 2011. It was insane what we got done in that time. And NASA was in the trenches with us, with Kathy driving the bus for quite a few weekend meetings."

Another factor was money. By the time of C2, the company had about 2,000 employees. SpaceX had won a lucrative NASA contract in late 2008 for the supply missions, worth $1.6 billion. But most of the payments would only come when the food and water started flying, and only a successful C2 would unlock this funding.

As the hardware teams hammered Dragon into shape, the mission operators were also hard at work. With Couluris in Hawthorne and Ridings in Houston, joint SpaceX–NASA teams learned to work together through dozens of simulations. These exercises ran the gamut from a nominal launch of Dragon through berthing with the space station to dealing with multiple problems and unlikely scenarios, such as Dragon being struck by debris or losing half of its Draco thrusters.

These sims allowed Couluris and Ridings, in particular, to form a tight bond. He directed a mission like he flew a plane—calm and in command. Ridings, in turn, brought actual experience operating spacecraft and the space station. These shared simulations fostered the trust necessary to handle stressful situations in future Dragon flights.

"We got to the point where we could hear and understand the inflection in each person's voice, and knew when they were okay with a decision, and when they needed more information," Couluris said. "It was like flying alongside someone in a cockpit, and it made for amazing teamwork."

But sometimes there were tensions when the hotshots from SpaceX came together with the buttoned-up engineers from NASA. SpaceX wanted to move fast and break things. NASA had a $100 billion space station and astronauts onboard to protect.

"They can be pretty aggressive about having ideas to fix things in real time," Ridings said. "It took time for us to come together. NASA is a team. SpaceX is a team. And we had to be a big team."

## If Dragon fails, the space station withers

NASA and SpaceX originally agreed to fly three Dragon demonstration flights, but SpaceX had already fallen behind schedule by the time C1 flew in December 2010. The second Dragon mission was meant to fly about one mile under the space station, perform some maneuvers, and then splash down. Only the third mission would berth.

One of the reasons NASA believed SpaceX needed a fly-under mission before berthing is that its positioning technology was unproven. Dragon used lidar, which stands for "laser imaging, detection, and ranging," as its primary location sensor. To determine distance, lidar bounced laser beams off the station, measuring the time it took the pulse to return. Dragon had two lidars and two thermal imaging cameras. These cameras took black-and-white images of the station and compared them to its known dimensions to determine distance and velocity. The thermal imagers were a backup, meant to cross-check the lidars.

As the date for the second demonstration flight slipped into 2012, however, these distance sensors started maturing. Couluris, along with guidance and navigation engineers Paul Forquera and Paul Wooster, agreed it might be possible to combine the objectives of the C2 and C3 missions into a single flight. It would be risky and ask an awful lot of the Dragon team. But it was the only way to make schedule.

Couluris first called Ridings, asking if this was something NASA might consider. Ridings discussed it with Lueders, and they both thought it could be done. But they needed a green light from Doug Cooke, the associate administrator. After Shotwell and Musk met with Cooke in the summer of 2011, NASA tentatively agreed to move ahead with the plan.

Couluris said financial pressures on SpaceX did not ultimately drive the decision to combine the missions, allowing C2 to berth with the station. Rather, it was the readiness of the proximity sensors and the general sense that, with a little luck, SpaceX could pull off the objectives of both missions in a single go. However, the decision did save the cost of a Dragon and a Falcon 9 rocket and bring the company closer to unlocking cargo flight payments.

Couluris acknowledged that SpaceX always operated under financial pressure. "Elon once told me, just like ingots of aluminum, or a pile of aluminum circuits, he had a small chest of gold coins that he could devote to this," Couluris said. "It was this natural resource that we needed to be responsible about. So we were scrappy."

The new plan for C2 meant that SpaceX needed to train astronauts on how to grab the vehicle. The space shuttle, as well as Russia's Soyuz and Progress spacecraft, docked directly with the station. But for the sake of simplicity, Dragon lacked the capability to autonomously control its flight all the way to a docking port. An astronaut on the station needed to reach out with the station's large robotic arm, grab Dragon, and move it to a docking port.

Astronauts had been visiting SpaceX for years to help guide the design of Dragon. Some of the first visitors were toured around by

Shotwell herself, or got the perk of driving Musk's Tesla up and down a side street near the factory. One of the people who frequently met with the astronauts was Laura Crabtree, who worked for Couluris. They had a lot of questions about the brash company.

"We were trying to ease their concerns," Crabtree said. "Everybody thought we were kind of space cowboys, and we wanted to show them that we were here to stay, and that we were doing things in the best and safest way that we knew how."

As the C2 flight got closer, Crabtree helped with the Dragon training. Among the astronauts who rotated through was a brilliant and charismatic engineer in his late fifties named Don Pettit. He ended up being the sole NASA astronaut onboard the station in May 2012 when Dragon came calling.

Back in Houston, among the astronaut corps and engineering staff, a lot of skepticism remained as the launch of C2 neared. Lueders, who by then had worked with SpaceX for more than half a dozen years, frequently encountered this in the halls at Johnson Space Center. Three or four times a week, someone would come up and say, "I'd hate to have your job." They just did not believe SpaceX could pull it off.

But Lueders understood that NASA had no choice. "This was our Hail Mary pass," she said.

NASA had spent the better part of a decade and a half assembling the International Space Station a few hundred miles above the surface of the Earth. It was a triumph of engineering and international diplomacy, as the United States worked closely with Russia, Europe, Japan, and Canada to operate the station. One of the main reasons NASA built the station was to study human health in zero gravity and conduct scientific research in the otherworldly environment. But the space shuttle's retirement blew a huge hole in those plans.

To conduct biological research, NASA had to send experiments up in freezers and other controlled conditions. So it had levied a requirement on SpaceX and Orbital Sciences to build powered cargo lockers into their spacecraft. The two other supply vehicles that NASA relied on, built by the European Space Agency and Japan's space agency, did not have this capability. And none of these vehicles, nor Russia's, could return science experiments to Earth. Only Dragon was designed with the ability to splash down into the ocean, be recovered, and be reused. So if Dragon did not work, the massive and costly station NASA built in orbit would become effectively useless for biological experiments.

"We had it on shuttle, but then it went away," Lueders said. "If we didn't get powered cargo up, and sample return down, you couldn't do any of the biological research. Maybe my colleagues were right, and SpaceX was going to fail. But what choice did we have?"

## Ridings to the rescue

A few days before the C2 launch, Couluris gathered all four shifts of his operations team in SpaceX's Mission Control Center. He told them he was proud of their hard work and believed they had taken every reasonable precaution for a safe flight. SpaceX had even gone so far as to prepare for the contingency of a large earthquake in Southern California, in which case Dragon's control would be passed to a backup team at Cape Canaveral.

During the flight, Couluris said the team would almost certainly experience things they had not simmed for. They probably were going to have difficulty sleeping the night before launch, like pilots before a critical sortie. No, he cautioned, they should not drink any alcohol or take medication to sleep. That would just make them groggy. But the most important message Couluris had for his team was simple: "You are ready. We've done everything we can."

The third Falcon 9 rocket launched at 3:44 AM local time in Florida, or shortly after midnight in California. As director of the Dragon program, Giger attended the launch before catching a flight back to Hawthorne once Dragon reached orbit. Along the way he pulled telemetry data using the plane's Wi-Fi network. Because this was Dragon's first prolonged spaceflight, SpaceX and NASA took it slow. The vehicle went through a long battery of tests in orbit and was not due to berth with the space station until the third day of the flight.

Giger and a few other Dragon engineers were so anxious, they did not plan to leave mission control for the duration of the flight to the station. Shotwell, ever the adult in the room, intuitively understood that the Dragon team would have to sleep sometime, so she rented a handful of vintage Star Waggons and put them in the Hawthorne parking lot. These were modified motor homes that Hollywood provided for movie stars to rest in between scenes, with couches and areas for makeup. SpaceX got castoffs from the movie industry, vintage models from the 1980s, complete with orange shag carpeting and lots of vinyl.

"We just did not want to go home in case something crazy happened," Giger said. "We were running on pure adrenaline. I think I slept maybe an hour between the launch and berthing attempt."

Altan watched the liftoff from Hawthorne and felt the buzz in the control room as Dragon deployed its solar arrays and everything went to plan. But anxiety gnawed inside Altan as his big test approached. He had responsibility for Dragon's avionics, which included the ability to communicate with the space station. Two years earlier space shuttle *Atlantis* had carried a briefcase-sized radio with a clunky name, the Commercial Orbital Transportation Services Ultra High Frequency Communication

Unit, to the station. SpaceX referred to it as CUCU, pronounced cuckoo. NASA did not particularly appreciate this acronym, but when SpaceX engineers proposed a few even more ribald options, the agency stuck with CUCU.

In orbit, astronauts wired the CUCU radio into the cabling that runs along the length of the station. These cables ran hundreds of feet out to a large antenna at the end of the station. What worried Altan is that although the CUCU radio emits a very strong signal, it gradually weakens as it propagates along the cables. By the time its signal is emitted from the station's antenna, it literally has the power of a Bluetooth device.

NASA required Dragon to initiate communication with the space station at a distance of seventeen miles, and Altan fretted that Dragon's two antennas would not be able to receive the weak signal so far away. If communications could not be established, the mission would be called off and Dragon sent home.

"I was sweating bullets the whole time that something I was in charge of was going to waive off Dragon," Atlan said.

He stayed up all night through the launch and into the next day while Dragon slowly drew nearer to the space station. But in the end, Altan need not have worried. Dragon picked up communications from the station at a ridiculously long distance of 250 miles. After Dragon locked on to the station, Altan let out a huge sigh of relief. Then he left mission control and crashed in one of the Star Waggons.

The mission went reasonably well for the first two days, before Dragon flew under the station for the first time, reaching within 8,200 feet. At this point Dragon would attempt to meet all of the objectives of the original C2 mission, including demonstrations of guidance and navigation systems. If Dragon passed these tests, it would be cleared for a berthing attempt the next day. However, as it flew under the station,

Dragon experienced more than half a dozen failures, including the loss of its GPS system for navigation.

All the hard work done between the C1 and C2 missions paid off. There had been no redundancy in any of the C1 spacecraft's systems. But now, Couluris's team could clear each of these single-event upsets by going to Dragon's backup systems. And so the spacecraft proceeded, ultimately flying a racetrack-like loop around the station before returning to a point several thousand feet below and clearing the way for a berthing attempt.

After a few hours of uneasy sleep, Couluris returned to his console on the evening of May 24. In Houston, Ridings and her team came into mission control a little later. Up until this point SpaceX retained authority over Dragon. But once the vehicle moved inside a distance of 1.5 miles from the station, the astronauts onboard could command it to abort for any reason, and Ridings had the final say. With this in mind, she gave the go for Dragon to start slowly climbing from its position below the space station to a distance of just 800 feet.

As it moved to this hold point, one of the spacecraft's two lidars started to return some glitchy data. NASA's mission rules allowed Dragon to continue approaching the station as long as one lidar functioned. Losing both meant an automatic abort. There was another problem as well. The two thermal imagers were supposed to cross-check the lidar data with, at most, a 10 percent error. But they were returning errors of up to 20 percent. Couluris and Ridings discussed the issue at length over their private loop, eventually agreeing to expand the acceptable error limit on the thermal imager readings to 20 percent. They believed that the errors would decrease as Dragon got closer to the station.

Precision mattered. Dragon and the space station were both hurtling around the Earth at a remarkable speed, about 17,000 mph. This is thirty times the speed that a passenger jet flies at cruise altitude. Relative to each other they were nearly stationary, but NASA was faced with a spacecraft that massed several tons and, if it went off course, could easily smash into the inhabited space station.

Dragon began climbing again, slowly. At a distance just inside 250 feet, the first lidar began to return false positives. At 200 feet both lidars indicated they were close to faulting.

Then lidar two outright failed.

Paul Wooster called for Dragon to retreat back to a safe distance. Couluris turned to the engineer in charge of navigation, Jeff Tooley. For critical spacecraft maneuvers both had to send a command, kind of like Captain Kirk and Spock both had to order a self-destruct for the starship *Enterprise* in the original Star Trek series. They issued the command, and Dragon began to back away. Couluris then called Ridings and explained what the team had done.

Nineteen seconds had elapsed since the lidar's failure. "Because we had practiced so much together, and the team worked so well together, we were able to execute that [maneuver] quickly," Couluris said. "We had to, because Dragon was really close to aborting."

With that, Dragon returned to a distance of 250 feet from the station, the point where the lidars had begun to send funky returns. This gave the SpaceX and NASA teams time to assess the problem, and they determined that lasers from the lidar were not just bouncing off dedicated reflectors on the station but also off the nearby Japanese module, Kibō, a phenomenon known as "dazzling." The navigation team suggested they reduce the area of the station the lidars were observing. With a smaller window, the Kibō module would be excluded. This involved a minor software change, however, and doing that in proximity to the station violated NASA's flight rules.

The decision was in Ridings's court.

As she deliberated in Houston, Couluris had other pressing matters. His team analyzed the amount of propellant left onboard Dragon, and they were not sure enough remained to make another go of it. That is, if Ridings did not allow Dragon to proceed that morning, Dragon might lack the fuel needed to make another attempt on the next day.

Couluris also worried about the vehicle's thrusters. As Dragon held its position a few hundred feet below the station, its orbital velocity was slightly faster. Over time it drifted ahead of the station. To counteract this drift, the Draco thrusters were firing on and off, in rapid succession, acting as brakes. This caused them to heat up rapidly. When they're full on, Dracos are self-cooling. But with rapid pulses the thrusters heat up, and then shut down before the cooling system comes on. Couluris and his team saw Dragon's thrusters nearing their redline temperatures, which would also have meant an automatic abort.

"It was excruciating watching the temperatures creep up, and we were getting super close to overheating," Giger said. "We really had to go, one way or the other, or the mission was done."

But Ridings needed time to assess the situation. Her job, as she saw it, was to get Dragon to the station safely. She did not want NASA to be the guardian at the gate, but she also had to protect the station. In the realm of NASA flight directors, a software change such as the SpaceX team was proposing rang the alarm bells. A few colleagues told Ridings that SpaceX should not be allowed to touch Dragon's software, lest this change introduce an error elsewhere.

Most of us wouldn't want to be burdened by the pressure to make such a high-stakes decision. "From my standpoint, as a flight director, this is what you train for," Ridings said. "This is the most fun you ever have, because you have a really difficult problem. And you've got to decide what to do."

So Holly Ridings decided.

While consulting with her team in Houston, she had also been talking with Couluris on that private loop. Ridings could tell by the tone in his voice that SpaceX felt confident in its solution. The location sensors had been developed by an engineer named Andrew Howard, and he had Couluris's trust. And after years of training together, Couluris had Ridings's trust.

She overrode the concerns of some members of her team and told Couluris to update the software.

After the software change, SpaceX reset both lidars and resumed climbing toward the next hold point at 100 feet from the station. Onboard the station, Don Pettit started positioning the robotic arm to grab Dragon. After holding at a distance of 100 feet, Dragon was cleared to move toward its final hold point, a mere thirty feet below the station. However, as the spacecraft passed through fifty-five feet, its second lidar started to dazzle off the docking ring.

Then it failed.

"We were so close," Couluris said. "We were just sitting there sweating because we were on one lidar, and two thermal imagers. If we lose the cross-check, or we lose the lidar, it's game over."

Dragon flew on, painfully and deliberately slowly, less than three inches per second. SpaceX reset the second lidar, but it failed again, just as the spacecraft reached its final hold point.

Then Pettit radioed down from the station that he was good to go, so SpaceX put Dragon into a free drift, meaning it would no longer fire its thrusters. Pettit would have about fifteen minutes to grab Dragon before it drifted out of range. As Dragon began to drift, its one remaining lidar started showing intermittent dazzling.

Kathy Lueders watched this drama play out from a room near mission control in Houston. She had heard the grumbling from some officials when Ridings decided to allow a software change, but she felt the flight director had made a great call. Now she sweated as Pettit guided the arm slowly toward Dragon, concerned the remaining lidar might fail at any moment.

"I was just like, really, Don, could you go any slower?" Lueders said. "Part of the slowness was me just wanting him to get it. He didn't really go slower than the simulations. But we were so close. We'd been through three days of working on the spacecraft issues. We're finally here. It was pretty gut-wrenching."

Then, at 8:56 AM local time in Houston, Pettit had it. He called down to mission control in Houston that he had caught a Dragon by the tail.

Lueders felt overcome by the sight of the spacecraft attached to the space station. It had been six years of pressing and pushing colleagues to go along with this crazy commercial idea, one that not many people at NASA thought would work. Now their baby Dragon had left its nest and spread its wings.

"You can't imagine how beautiful it is when those spacecraft show up," she said. "It's just amazing. The way the spacecraft come in, they're just out of the dark. There's this tiny little light you see coming up, and it's just powering its way in. I'm getting emotional right now thinking about it."

In Hawthorne, they were out of their minds with delight. Before the mission, most of the Dragon team had given the spacecraft a 50 percent chance, at best, of reaching the station. The most likely scenario had Dragon getting pretty close, gathering a lot of data, but ultimately having to turn back. But they had gone all the way. Musk walked over to Couluris, shook his hand, and thanked him for the "incredible" job. That was high praise from Musk, not one for flattery.

As an operations planner, Laura Crabtree sat at the back of the room. Due to the hours-long delays in Dragon's capture, she had been coordinating with NASA's ops planner on modifications to the space station's schedule. As her colleagues leaped to their feet, hugging and high-fiving, Crabtree sat back in her chair and cried.

"All of these things we had planned and trained for had come to fruition," she said. "They were tears of relief and joy. We had done something at that moment that only countries had done. The gravity of what we had accomplished started to hit us."

## "The most significant thing I've done."

An assistant to Musk and Shotwell, Alyssa Sager, had ordered several hundred bottles of champagne. They were lined up just outside Mission Control. The bubbly flowed as those in mission control joined their colleagues on the factory floor. Giger, Couluris, Altan, and the other Dragon operators were delirious with joy and fatigue as they plowed through the champagne.

A couple of hours after Pettit had grabbed Dragon and moved it to a berth on the station, NASA arranged a news conference. Ridings and Suffredini joined from Houston. After opening remarks from NASA officials in Houston, the video cut to Hawthorne, where Musk sat next to Lindenmoyer, who had a grin on his face from ear to ear.

"I don't have words to describe the excitement and elation we have here at SpaceX," Musk said. "There's so much that could have gone wrong, and it went right. We were able to overcome some last-minute issues at NASA mission control and SpaceX mission control. It's just a fantastic day. This really is going to be recognized as a significant historical step forward in space travel. And hopefully the first of many."

Musk and Lindenmoyer sat just outside Mission Control, in front of the C1 Dragon that had been recovered from the ocean a year and a half earlier. Facing them were a few hundred employees standing in a large crescent. Later in his remarks, when Musk praised the hard work by his entire team, the crowd responded with an extended cheer, chanting,

"Elon! Elon! Elon!" It was entirely genuine and heartfelt. Musk paused for a moment and hugged Lindenmoyer. Then he thanked NASA for believing in SpaceX.

A few other senior Dragon officials were perched on the floor nearby during the news conference. They were so caught up in the moment, so full of happiness and relief that they briefly considered rushing over to Musk and hugging him on television. "I'd say, at the last second, cooler heads prevailed," Couluris said.

After Dragon berthed that morning, it was time for the three-person space station crew to sleep. This gave the SpaceXers who had been on pins and needles for the last three days, who had worked overnight, about twelve free hours before they would need to be back in mission control for the hatch opening. Exhausted, they went home and slept.

Just kidding.

The Dragon crew ventured out in the middle of the day on a Friday, in Los Angeles, and partied. Giger, whose family came from Switzerland, found a Swiss restaurant on the Venice Beach Boardwalk called "On the Waterfront." It served bratwurst sausage and large steins of Bavarian beer. Giger had not been to the beach in more than a year. Now he watched the surf and chilled with his friends and colleagues, buzzing from the beer as well as a deeply satisfying sense of accomplishment.

The next day Pettit; André Kuipers, the European astronaut onboard the station; and Russian cosmonaut Oleg Kononenko opened the hatch to Dragon and found everything was well. Pettit quipped that Dragon smelled great inside, like a brand-new car. Dragon spent six days berthed to the space station before Pettit released the spacecraft, and it burned its Dracos for home. Dragon landed in the Pacific Ocean, off the Baja Peninsula, on May 31, a little more than nine days after it lifted off.

Less than half a year later, SpaceX flew its first operational mission to the station, dubbed Commercial Resupply Services-1. When astronaut Suni Williams opened Dragon, she found a treat inside its powered lockers. Four hours before liftoff, SpaceX had loaded a carton of ice cream from Texas—Blue Bell vanilla with chocolate swirl. There were also plenty of science experiments. Thanks to Dragon, the International Space Station could deliver on its promise of groundbreaking scientific research.

For Couluris, the successful completion of the C2 mission validated his decision to leave JetBlue and follow his father's footsteps into the space industry. He would go on to work for the other space billionaires, serving as head of launch for Richard Branson at Virgin Orbit and leading lunar programs for Jeff Bezos at Blue Origin. What most impressed Couluris about Musk was his willingness to take chances in order to make great leaps. And SpaceX had done that by combining C2 and C3 into a single mission, meeting NASA's requirements in record time, and then pulling the whole thing off.

"The mission was life-affirming for me," he said. "All that time being a pilot and flying those missions, and then joining SpaceX and working through all of those simulations, was for that single moment in time."

Holly Ridings, too, walked away from Dragon's first mission to the space station knowing a lot more about herself and with more confidence in her abilities. Her bosses at NASA noticed as well. Thanks in part to her bold decision-making during Dragon's ascent to the station, she was promoted to become NASA's first female Chief Flight Director in 2018. Today, she is helping lead the space agency in its quest back to the Moon half a century after the Apollo Program.

For all of that, the Dragon mission still matters most.

"At the time I really didn't realize the impact," she said. "But then you roll forward eleven years. For me personally, with my belief in peaceful human exploration, and that we're all in this together, this was absolutely the most significant thing I've done in my career."

| 6 |

# AIN'T COMING BACK

*January 2012*

Vandenberg Air Force Base, California

When SpaceX began flying missions for NASA, its cachet as a destination for young engineers increased. A slew of eager new applicants, like a former Alaskan bush pilot named Benjamin Kellie, began applying. Although his engineering studies at The Ohio State University had focused mostly on alternative energy projects like wind turbines, rockets and spaceships sounded like a hell of a lot of fun to Kellie.

After he passed an initial round of telephone interviews, the SpaceX recruiter invited Kellie to Hawthorne to meet with senior engineers in January 2012. He worked out travel arrangements to arrive on a Sunday, ahead of meetings early Monday. Shortly after he landed in Los Angeles, however, Kellie's plans changed when his mobile phone buzzed. The director of SpaceX's launch operations at Vandenberg Air Force Base, Lee Rosen, wanted to know when Kellie planned to arrive there, 150 miles north of the city.

"I didn't know this guy, and he was saying I should come to his house," Kellie said.

Knowing he faced a series of grueling interviews at the company's factory the next day, Kellie had planned to take it easy on Sunday evening. But Rosen wanted to meet him in person and asked Kelly to make a three-hour drive north for an impromptu dinner. Kellie figured he better go with the flow, so a few minutes later he picked up his rental car and submerged himself into the endless river of LA freeway traffic.

A long-time Air Force veteran, Rosen had traded his commission for a SpaceX badge a year earlier to lead the company's buildup of a second Falcon 9 launch pad. Set amid the rugged hills climbing away from the Pacific coast, Vandenberg offered an ideal site for polar launches. Instead of following an orbit from west to east, polar satellites fly—as the name suggests—over the planet's poles. This turns out to be a useful vantage point for observing the Earth. As the planet spins below, a satellite looping around the poles can observe every part of the world over the course of a full day.

Rosen had asked Zach Dunn to manage development of the launch pad, and there were only a handful of other engineers and technicians working on-site at the time. But as pressure mounted to prepare the launch pad, they needed new blood.

Kellie met this small group at Rosen's house, where they talked about a good many different things, rockets and otherwise. As they talked, they drank homemade wine. And soon they were playing the *Dance Dance Revolution* video game. Eventually the wine gave out, and Rosen's wife, Dorothea, brought Kellie some coffee. This helped him sober up for the long drive back to Los Angeles.

He got back to his hotel around 4 AM. After tossing and turning for a few hours, Kellie dragged himself into the SpaceX offices, tired and hung over. This was decidedly *not* the best preparation for a day of interviews. During this process a candidate is typically sequestered in a small meeting room along the back wall of SpaceX's main engineering office, on the first floor of its Hawthorne headquarters. Each of these rooms is named after a spaceflight hero, such as Buzz Aldrin. Over the course of a few hours, a candidate usually meets with four or more different SpaceXers who rotate through and ask questions to size up a potential hire. The discussions can become energetic or, with white boards on the wall, require an applicant to work through equations or other technical problems.

Fortunately for the bleary Kellie, he did not have to deal with a full court press of questions. Rather, when Vice President of Launch Tim Buzza met with Kellie, Buzza asked no questions at all. Then managing both launch operations and Falcon 9 production efforts, the frazzled Buzza had little time for long interviews. Instead, Buzza turned to Kenton Lucas, who was leading Kellie's interview process. Lucas shrugged and said the Vandy guys liked his grit and creativity. That was good enough for Buzza. Kellie and Lucas had lunch and wandered around the factory. By the time his plane arrived back in Columbus, Ohio, that afternoon, Kellie had an offer in his inbox to help build the Vandenberg launch pad.

"Later on we would joke that my on-site interview was getting drunk and playing *DDR*," Kellie said.

## The Great LOX Boil-Off of 2013

SpaceX had attempted to launch from Vandenberg nearly a decade earlier. Partly due to its proximity to SpaceX's factory, Musk chose the Air Force facility for the Falcon 1. After a small launch team built a pad, they completed the company's first static fire test in May 2005. But then the Air Force got cold feet. The military had sunk $200 million into retrofitting a launch pad, Space Launch Complex 3, for Lockheed to launch its new Atlas V rocket. And near SpaceX's Falcon 1 facilities, a mighty Titan IV rocket stood on a launch pad with a $1 billion spy satellite for the National Reconnaissance Office.

So when push came to shove, it was SpaceX that got shoved. Air Force officials told Musk he would have to wait several months, if not longer, for the Titan IV to launch before he could attempt an experimental flight of the Falcon 1. Seething, Musk abandoned Vandenberg and threw SpaceX's remaining resources into launching from the distant Kwajalein Atoll in the Central Pacific. It had been a desperate throw of the dice that almost killed SpaceX.

Six years later the Titan IV was retired, and SpaceX had proven itself to be a responsible operator with the Falcon 9. The Air Force welcomed SpaceX back and agreed to lease the old Titan launch pad, Space Launch Complex 4 East, to the new guys. Rosen's hiring helped smooth the way. Having served as commander of the 4th Space Launch Squadron at Vandenberg before, Rosen knew how to speak the Air Force's language and meet their requirements.

There was so much copper in the old launch tower that a contractor paid SpaceX $1 million for the opportunity to demolish the Titan facility. After this, SpaceX prepared the site for its own needs by pouring concrete, building a processing hangar, and installing large (often salvaged) tanks for storing liquid oxygen and kerosene propellant.

Rosen's team raced against the engineers back in Hawthorne, who were hard at work hammering out a bigger and better Falcon 9 rocket. Whereas Falcon 9 version 1.0 could lift only the Dragon spacecraft, version 1.1 would be able to fly commercial satellites as well, opening the era of nongovernment launches on Falcon 9. As this larger rocket would make its debut in California, the company's future ran through Vandenberg.

"We were always under the gun," Dunn said. "There was always the threat of the rocket being ready. The rocket was always going to be ready in six months. Of course, the rocket didn't show up until the middle of 2013. But from 2011 onward the rocket was always going to be there in six months."

SpaceX was a decade old by this time, but it still acted very much as a startup—a whirling vortex of activity and chaos and urgency. Like moving into a house where the paint on the walls was not yet dry, the first Falcon 9 1.1's arrival in Vandenberg in the summer of 2013 included lots of wet paint, both on the rocket and the launch pad.

Following tests to ensure the fit of the Falcon 9 with the transporter that carried it to the launch pad and raised it vertical, the focus turned toward a static fire test. This would stress both the new rocket's propulsion system as well as the recently built ground systems that fuel and support the vehicle just prior to liftoff. By late August, the launch team, led by Buzza, decamped from Los Angeles to lead this critical test.

After several fits and starts, the rocket finally reached the last seconds of the countdown, when the turbopumps inside the engine begin to spin up prior to igniting oxygen and kerosene in the combustion chamber. However, an operator mistakenly left helium flowing into the engines prior to ignition, and this forced an abort of the test. Per their procedures, the engineers opened valves on the rocket to drain propellants and recycle operations for another go at the static firing. However, when the fill-and-drain valve leading to the liquid oxygen tank on the second stage was commanded to open, nothing happened. It was stuck.

If you have ever watched a rocket launch, chances are you have noticed white, puffy clouds moving away from the vehicle during the final minutes of the countdown. This is liquid oxygen "boiling" off from the storage tanks inside the vehicle. Most rockets use liquid oxygen because it is the most potent oxidizer for the combustion of a propellant, creating an efficient reaction. Essentially, it produces more bang for the buck than other oxidizers. The challenge of liquid oxygen is that it boils at an extremely cold temperature. Whereas water boils at 212 degrees Fahrenheit (100 degrees Celsius), one of its main constituents, oxygen, boils at –297.3 degrees Fahrenheit (–182.9 degrees Celsius). Chemistry, man, it's a gas.

In the control room near the rocket, Zach Dunn's heart sank when he realized the fill-and-drain valve had gotten stuck. He immediately grasped that the launch team had no option but to wait it out. The LOX inside the Falcon 9 rocket's upper stage would eventually boil off, of course. But it would not happen fast because there was so much of it.

"Vandenberg is foggy and 45 degrees, you know, so it's not going to boil off quickly," Dunn said. "We literally had to wait for three days."

It was a painful wait. While the kerosene had been offloaded from the rocket, the SpaceXers could not get close to the pad. The stubborn Air Force officials, who were responsible for range safety, would not let engineers and technicians climb onto the rocket and dislodge the valve. Some crazy ideas were considered by the increasingly desperate launch team. Perhaps the stage could be shot with a rifle? After the LOX drained through the resulting hole, they could patch up the stage and go. But upon reflection, the idea of sending an energetic projectile through volatile liquid oxygen seemed like it might not be the best idea.

And so they waited. While the liquid oxygen boiled away steadily, there was a lot of it, nearly 10,000 gallons. By the third day Dunn neared a breaking point in the control room, watching the level of LOX in the second stage tank slowly fall. Nearby another engineer, Ricky Lim, had fallen into a routine. Every fifteen minutes Lim would click the switch that commanded the valve to open. When it didn't, he would play Led Zeppelin's classic 1969 track "How Many More Times" on the sound system. As Robert Plant plaintively sang the chorus and Jimmy Page played a mournful, bluesy guitar, Dunn would mumble to himself, "Please don't be water, please don't be water, please don't be water." The track stretched on for nearly nine minutes. After it ended, it was almost time to test the valve again, to see if it would open this time.

Dunn's fear was that water had somehow gotten lodged into the valve, causing the stuck valve. After the valve had closed, with the second stage tank full of lox, this water would have then frozen due to its proximity to liquid oxygen. Dunn managed development of all of the fluid, electrical control, and high-powered electrical systems at SpaceX's Vandenberg launch site, so this would be his fault.

"I was deathly afraid that it would be water in the actuation line because it was fed from the ground side that I was responsible for," Dunn said. "I really didn't want it to be my fault that we just lost three or four days."

An actuator is the mechanism that physically opens the valve. This is achieved by a dry nitrogen gas moving up a quarter-inch line to the actuator. As part of their preparations for the static fire test, Dunn's team had cleaned and dried this line, as well as other actuators and valves on the ground systems, to ensure there was no water in the line for the very reason that it would freeze.

There were other possibilities, of course. Through some unknown action, for example, debris could have lodged the valve closed. The acid test for Dunn's waterlogged actuator would come when the temperature of the valve reached 32 degrees Fahrenheit (0 degrees Celsius), at which point any ice would melt. Another sensor in the control room measured this valve temperature—something else for Dunn to anxiously track as Robert Plant crooned on.

"Sure as shit, it gets to 32 degrees and the valve flies open," Dunn said. "After they lowered the rocket they took the actuation line off the stage, and this gunky, terrible brown water poured out of it. It may have been from the cleaning process, or something else. That was just super embarrassing and a big setback."

## How to play schedule chicken

Tim Buzza endured his own, separate hell during the Great LOX Boil-Off, as it came to be known. One of Musk's earliest hires back in 2002, Buzza became one of the company's most respected and beloved employees. He had a brilliant mind and a big heart, and, having grown up in a blue-collar family in Pennsylvania steel country, he was accustomed to hard work. By the fall of 2013, Buzza had put in eleven arduous years at SpaceX, leading dozens of static fire and launch countdowns. But the events of the boiloff, and Musk's reaction, damaged their relationship beyond repair.

"That was the end of SpaceX for me," Buzza said.

Most of those eleven years were difficult but ultimately rewarding years for Buzza. After three failures, Buzza's launch team delivered success with the Falcon 1 rocket in September 2008. And then, against the odds, Buzza had turned around and put the very first Falcon 9 into orbit less than two years later. Still, Musk asked for more. In addition to managing launch operations for the Falcon 9, Musk also asked Buzza to oversee production of the rocket.

Model 1.0 of the Falcon 9 booster had been a development version, but the 1.1 iteration was intended for mass production. Musk liked to set straightforward, but ultra-ambitious, goals for his launch team. For example, he wanted Buzza to be able to roll a rocket out of its hangar and launch in an hour. He wanted to be able to launch one Falcon 9 rocket a week. And now he dropped a new target on Buzza for annual Falcon production.

"Elon was always able to come up with simple numbers," Buzza said. "Our new mantra was forty cores, twenty second stages, fourteen fairings, and six Dragons, and it all had to be done in Hawthorne," Buzza said. "We had to do this with 1,000 employees in a factory with half a million square feet."

To unpack this a bit, Musk was asking his team to rapidly create the capability to launch twenty rockets a year. (At the time, SpaceX averaged a single launch per calendar year.) The forty cores referred to both the single-core Falcon 9 rocket as well as Musk's vision for a "Falcon Heavy," which would consist of three Falcon 9 cores strapped together and provide substantially more lift to orbit. Accordingly, he envisioned ten annual launches of the Falcon 9, and ten of the Falcon Heavy.

If you're thinking this sounds like a great argument for reusing the first stage cores of the Falcon 9 rocket, you are not alone. "I thought reuse was viable," Buzza said. "I cheerleaded it hard. I didn't want to run production of forty cores."

As Buzza set about ramping up production, other departments worked on various improvements to the Falcon 9. The propulsion team upgraded from the Merlin 1C to the Merlin 1D engine, substantially increasing the engine's thrust while maintaining its weight. They also introduced a new process to manufacture the thrust chamber, the heart of an engine where propellants enter, combust, and produce the thrust that lifts a rocket.

This new manufacturing process was necessary after an engine issue on the fourth flight of the Falcon 9 rocket, which carried the first operational cargo supply mission for NASA in October 2012. During the ascent to orbit, one of the rocket's nine Merlin 1C engines failed after seventy-nine seconds. While the vehicle survived this engine failure, and Dragon delivered its Blue Bell ice cream to astronauts waiting onboard the space station, a secondary payload consisting of a small

ORBCOMM communications satellite did not reach its intended orbit. Musk tapped Hans Koenigsmann to lead the investigation, and he worked closely with Kevin Miller. After a couple of months, the pair were fairly certain the engine failed due to a poorly manufactured thrust chamber. SpaceX had to stand down for nearly five months as the investigation proceeded.

"The engine issue stopped us dead in the tracks with NASA," Koenigsmann said.

For the next cargo mission, the interim fix was to carefully examine each thrust chamber through a process called shearography to ensure there were no defects. As part of upgrading to the Merlin 1D engine, SpaceX had already designed and tested a brazed thrust chamber that moved away from plating the nickel-cobalt alloy onto the chamber. This killed the problem entirely.

Another major change occurred at the base of the vehicle. The original Falcon 9's engines were arranged in a tic-tac-toe pattern, and this left a massive space between the bottom of the propellant tanks and the top of the engine mounts. It was so roomy that engineers jokingly referred to this area as the "dance floor." This wasted space equated to extraneous mass, and Musk was determined to find a better design. He and the structures team considered two options, the "Coke bottle" and "Octaweb." The Coke bottle would effectively mount the engines almost directly onto the lower pressure dome of the propellant tank. But Musk ultimately preferred the Octaweb, which had one engine in the center surrounded by the other eight in a circle. This allowed room for sheets of aluminum between each Merlin engine, creating individual bays, protecting the other engines in case of one's failure.

Musk made this decision in February 2011. After their meeting, the director of structures, Mark Juncosa, put together a team to work on the Octaweb's design and construction. He picked Sam Stults to lead the project, and then the pair walked over to the desk of Robb Kulin. He was a materials science engineer who had studied bone fracturing to earn his doctorate degree. It was Kulin's third day at SpaceX. "Duderino, we're gonna rock your world," Juncosa told Kulin. "There's this new thing called the Octaweb, and you're now working on that."

The Octaweb exemplified the shift to version 1.1, which entailed a complete redesign of the rocket. Over the course of more than two years, the vehicle was rebuilt from the ground up, with an eye not just on performance, but on process as well. Musk forced his engineers to consider operations, optimizing the design for ease of building, transporting, and launching rockets. It was clear from the outset that, starting with this version, the Falcon 9 rocket was not just going to be another medium-lift vehicle.

"That set the foundation for the Falcon 9's future success," said Kulin, who would eventually become director of Flight Reliability. "The lessons learned testing and launching it for the first time, and the next few rockets, drove in numerous small reusability and servicing tweaks that live on today in the design and allow the Falcon 9 to be both reusable and processed as fast as it is."

As part of the buildup to the new rocket, the structures department had to develop a composite fairing, which sits atop the rocket and protects a satellite and makes the rocket lighter overall. And the avionics team made the rocket's electronics fully redundant, while radically shrinking the size and weight of computers onboard the vehicle.

Musk, as ever, pushed his teams faster and faster. In meetings about the rocket's flight computers, to emphasize the need for miniaturization, Musk would toss his mobile phone on the table and say, "My iPhone can do more than your avionics suite on the rocket." As usual with development projects, each team faced difficulties bringing their new hardware online.

"We played schedule chicken," Buzza said. "Whoever was the furthest behind would take the most heat, and all of us would tuck behind that as a cover. We all took our turns being the long pole. I would go and thank Mueller when he would tell Elon the engine was delayed. And they would thank me when the launch site was delayed."

Building a payload fairing was one of version 1.1's biggest challenges. A fairing is the bulbous enclosure at the top of the rocket that protects the payload during ascent, and then opens and falls away before a satellite is released. The dominant industry player at the time was a Switzerland-based company called Contraves Space, which later became RUAG. It built fairings for United Launch Alliance and its Atlas V rocket, selling SpaceX's competitor a comparably sized fairing for about $5 million. RUAG offered SpaceX a similar price for a Falcon 9 fairing. However, Musk believed that by building the fairing in-house SpaceX could, as usual, save a large chunk of money.

As part of Musk's plan to profitably sell a Falcon 9 launch for $60 million, he was counting on the fairing costing about $1 million. Buzza, however, began to have concerns when the production team started manufacturing the first fairings. It proved to be a labor-intensive process. The basic idea was similar to papier-mâché, in which wet pieces of newspaper are pasted onto a balloon to form a structure. For the fairing, SpaceX used sheets of carbon-composite fabric. A laser tool would indicate where to place a particular piece of the fabric, and a technician would lay it out by hand. As Buzza watched this work over the course of about half a year, the best technicians reached a point where perhaps 100 of these pieces might be set down during a day. But thousands of pieces had to be placed on each fairing half. Then the fairings had to be cured in a large oven.

As part of his production plans, Buzza ran the numbers. To meet Musk's goals, SpaceX would need to purchase a second oven. And based on the labor and materials, he did not believe a fairing could be built for $1 million. The final cost, he estimated, was probably closer to $4 million. This disparity between Musk's expectations and reality came to a head on a Sunday in 2012, the day before the first fairing oven was due to be delivered to SpaceX's factory. Buzza needed to know where to put it, so he telephoned Musk, who in turn called a meeting at the factory later that day.

Another key figure at the meeting was the structures director, Mark Juncosa. He had joined SpaceX in 2005 after graduating from Cornell University with a degree in economics. He possessed an amazing amount of energy and a distinctive, unruly mop of brown hair. Although he has limited public visibility, Juncosa has played a hugely consequential role in the company's Falcon 9 and Starship years. He intuitively understands what Musk wants, and attacks whatever task is in front of him with all guns blazing. When a program fell behind, Musk dispatched Juncosa to fix it, and Juncosa frequently steered large programs such as Starlink and Starship through troubled waters. Above almost everything else, Musk values the ability to get difficult shit done in creative ways, and Juncosa does this better than anyone.

"Mark Juncosa is the most important technical person at SpaceX other than Elon, and he just flies completely under the radar," said Abhi Tripathi, who worked at SpaceX for a decade before leaving in 2020. "If you've ever watched *Game of Thrones*, he's the King's Hand. He is the right-hand man and most trusted senior advisor."

By the time of the Black Sunday meeting, as it would become known, Juncosa had taken over running the structures department from Chris Thompson. Unfortunately, before the meeting, Buzza and Juncosa lacked the time to coordinate what they planned to say. This had simply been lost in the pedal-to-the-metal race toward the launch pad. After Musk called them together on a Sunday, he asked about the fairing's production. Buzza launched into a presentation about the time needed to produce each fairing from a labor standpoint, saying it would take about 5,000 hours, or more than half a year.

"Before we even got to the cost charts, Elon calculated that it would be about $2 million per fairing half," Buzza said. "It was just a microsecond. He's that quick with numbers. We didn't even have to show him the charts. And he said, 'Wait wait wait, this is not what is going to happen.'"

The meeting deteriorated from there, as Juncosa backed Musk's view that the fairing only needed one thousand production hours and should cost $1 million, not $4 million. The disagreement highlights the ambitious cost targets Musk wanted to hit even as he asked his teams to move at breakneck speed. With the Falcon 9 payload fairing, they ended up missing. Ultimately, a fairing would cost SpaceX nearly $6 million to make. Even though it brought the manufacturing process in-house, SpaceX was unable to change the basic economics of fairing production—if you're just laying out composites by hand like everyone else, you haven't really changed the game. The high cost ultimately led SpaceX to start recovering fairing halves and reusing them.

In March 2013 SpaceX launched its second operational supply mission for NASA, formally known as Commercial Resupply Services-2, or CRS-2. This was the final flight of the Falcon 1.0 rocket, and pressure ratcheted up on both the rocket development and launch site teams. Buzza felt pretty good about how the company was meeting its schedule and cost targets on most projects, but a series of disagreements like the Black Sunday fairing meeting had begun to fracture his once tight bond with Musk. These cracks ruptured during, and immediately after, the LOX boiloff incident.

The three-day wait agitated Musk, and he periodically expressed his impatience with Buzza by telephone. Musk thought the liquid oxygen should boil off overnight, but foggy Vandenberg was not Cape

Canaveral with its tropical climate and plentiful sunshine. Buzza said the real pain point came after the LOX finally did boil off. SpaceX needed to replace the fill-and-drain valve on the second stage, but it was buried deep inside the guts of the rocket. This necessitated lowering the rocket from its vertical position to a horizontal one, where it would be safer to get inside the Falcon 9.

This created another problem, however. Rockets like to exist under pressure. Take a can of Coke. When sealed, it is effectively pressurized and fairly robust. When opened, however, the can's internal pressure is lost and it can easily be crushed. A rocket's structure and propellant tanks operate under a similar principle. Therefore, during rollout to the launch pad, and once there, the second stage was pressurized with inert nitrogen gas. It remained so during the great boiloff. But before the valve could be worked on, the stage had to be depressurized. The concern was that if it was horizontal, the structure might buckle due to the payload at the end of the horizontal rocket, which would apply a downward force like an apple at the end of a tree branch.

There was some debate about whether this would actually occur. The payload in question, a relatively small Canadian satellite named CASSIOPE with a mass of 1,100 pounds, was unlikely to be heavy enough to cause significant buckling. Nevertheless, the chief of structures, Juncosa, had said Falcon 9 should be pressurized during rollout, and now he was anxious about it buckling when the pressure was taken off. He and Buzza worked together to improvise a solution, which entailed looping a belly band around the upper end of the rocket and lifting it with a crane to provide enough support to prevent any bending. In a worst-case scenario, the payload would fall off, releasing a dangerous brew of hypergolic propellants and causing a toxic explosion.

Musk had been otherwise engaged while the operation proceeded, but as it neared its conclusion, he checked a video monitor showing activities at Vandenberg. The view of a crane bending over the rocket enraged him.

"He called me and just ripped me," Buzza said. "He was furious. He was absolutely screaming at me. It got very aggressive, to the point where it felt super threatening."

What angered Musk is that, much earlier in planning for Falcon 9 version 1.1 operations, he had decided that only missions with the heaviest payloads were susceptible to buckling. He did not understand why his engineers at Vandenberg were wasting time with a crane and a belly band for the comparatively light Canadian satellite. Musk had one goal at that moment, and it was to get that damn rocket to orbit. Anything else was a superfluous waste of time—time that he did not have, or want to pay for.

So he unloaded on Buzza. The long-time launch director tried to explain that he was following the plan set out by Juncosa. He tried to tell Musk that the operation was just minutes from being completed and had not really cost much, if any, time. It was to no avail.

"That was the exact day that I decided that I was going to leave SpaceX," Buzza said. "But I had committed myself to get through this launch, and one more at the Cape."

He soldiered on for a little while longer.

## "What the hell are you so happy about?"

The Great LOX Boil-Off of 2013 did not end the travails for the Vandenberg launch team. A little more than a week later, on September 7, a calamity shook the California site that could have set the launch attempt back months.

By then SpaceX had successfully static-fired the rocket. Only a few major tests remained before a launch could be attempted. Rosen and

Dunn kept putting off one of these tests, of the water deluge system. This is because the test would shut down operations on the launch pad for a prolonged period, at a time when so much other work needed to be done.

A rocket launch produces a tremendous amount of noise and vibrations known as acoustic energy. As gas exits the engine nozzles, it exceeds the speed of sound and creates massive shockwaves. Noise levels at the instant of launch approach nearly 200 decibels. To help put this into context, a jackhammer is 100 decibels. The deck of an aircraft carrier averages 140 decibels. So the action at the ass end of a rocket is pretty damn intense. The problem is that this energy flows into the launch pad and can be reflected up the rocket, with its fairly fragile structure, and even to the delicate satellite at the top of the vehicle.

The solution for this is a water deluge system, which pumps a large volume of water into the base of the rocket and the area where exhaust is produced. This dampens the sound and reduces heat damage to the pad. Often, such sound suppression systems work with a flame trench that channels rocket exhaust away from the vehicle on the pad.

After Ben Kellie hired on at Vandenberg, he inherited responsibility for the water deluge system. It was fed by a giant 270,000-gallon water tank about 1,000 feet higher than the launch site. This is a lot of water. The water stored in this tank could almost fill an Olympic-sized swimming pool halfway. It flowed downhill in a three-foot pipe, and then as it neared the launch pad this water main was diverted into six different pipes, each of which led to a nozzle that sprayed water onto the launch pad.

SpaceX had installed butterfly valves leading to each of the six different pipes and tested them at lower pressure. One of the biggest concerns of a high-pressure water test was shutting down the flow through these valves too quickly. Water running downhill, in a three-foot pipe, chugs along like a freight train. Imagine running at full speed, and someone shuts a door in your face. With water the effect is actually even a little bit worse because there is some elasticity in compressible fluids, which means that there is a spring compression force that blasts backward. The technical term for this pressure shock in a pipe is "water hammer."

And that's what happened on September 7, 2013, as the Vandenberg team worked frantically to reach the point at which they could launch. As part of the water deluge test plan, at its conclusion, Kellie had written instructions to close each valve individually. He intended this to mean one at a time, waiting for each valve to close before closing the next. But the controller, who was being trained, interpreted this as needing to close the valves as fast as possible. This over-torqued the butterfly valves, creating the dreaded water hammer.

The water main busted.

The launch site took on the appearance of the golf course in the movie *Caddyshack*, when Bill Murray pumps groundwater to flush out a gopher. Water came rushing downhill. It bubbled up beneath SpaceX's main building on-site. As it emerged from every crack and orifice on the launch pad, water flooded the surrounding area. The main road leading from the launch site down to the rocket's propellant tanks washed out.

"I thought we were fucked," Dunn said. "I thought this would set us back months and months."

We should step back for a moment to understand how excruciating this must have been for Kellie, Dunn, and others at Vandenberg. Kellie had been working for months, without a single day off, in the final Sisyphean push toward launch. Technically, he worked the "day shift" on-site, which lasted from 3 AM to 3 PM. But as lead engineer for the pad, out of a sense of responsibility and ownership, this meant he arrived on-site at about 1 AM and worked until 5 PM. Makeshift beds were set up in the Vandenberg offices for Kellie and other engineers to sleep in because they were too exhausted to drive home. After months of this, Kellie just wanted the ordeal to end, and now his system had flooded the entire launch site.

"You're just exhausted, and you're wondering, 'When is this going to end?'" Kellie said. "And I just crumbled. Not because I made a mistake, which was bad enough. It was like, Christ, now launch is gonna be even further. I needed this fucking rocket to go to space. I used to cuss at the rocket. I used to just look at it and say, 'Go to fucking space.' I would mumble at it, just like that. We were out of our minds."

Kellie felt like he had to own this mistake. After a little more than a year on the job, he would offer to resign for this monumental screwup. He walked down from the launch pad, on the road leading to large tanks where the Falcon 9's kerosene fuel was stored. He found Dunn, Rosen, and Buzza gloomily staring into a sinkhole that was about ten feet across. Kellie said he had screwed up and indicated his willingness to resign over his mistake.

"Zach went apeshit on me," Kellie said. "Basically, he told me, 'Oh, you're going to quit and just leave us here to clean it up on our own? That's your plan? We did not hire you for everything to go right. We hired you for when things are messed up. So feel bad for yourself for about ten more seconds, and then hitch up your boots and let's fucking go.'"

And so they went. After clearing their heads, Kellie, Dunn, and Rosen called on local contractors who specialized in earth work. By midnight backhoes and other heavy machinery were in place to clean up the mess. To their surprise, the cleanup only took a couple of days.

Three weeks after the great deluge, the Falcon 9 rocket was ready. Space Launch Complex 4 East was ready. And Ben Kellie was damn well ready. In addition to his other responsibilities, he coordinated the pad technicians, and during the wee hours of September 29 they walked around the rocket for a final time, making last-minute checks. It was a rare, clear night at Vandenberg, where the cold Pacific waters so frequently produce coastal fog. Kellie marveled at the stars shining above.

Just before the "Red Team" left the pad that morning, the lead technician, Russ Chai, instituted a new launch tradition for SpaceX. He smashed two pies, lemon meringue and chocolate cream, into Kellie's face.

Then they decamped to a small cinder block building a few thousand feet away. This was located as close as one could legally stand to the rocket, at the quantity-distance line, which essentially meant it was just beyond the blast radius. Kellie and Chai donned firefighter gear, standing side by side as members of the Red Team responsible for securing the pad after launch. They listened to the countdown through a tinny loudspeaker.

Inside the flight control room, the launch team worked no technical problems with the rocket or its ground systems. But they were closely monitoring upper-level winds, which were very near the limit of what the Falcon 9 could tolerate. The issue was not so much the absolute wind speeds at higher altitude, but rather shear. This occurs when the direction and velocity of winds changes rapidly as a rocket ascends through the atmosphere. High amounts of shear can damage or even destroy the structure of a rocket. Musk succinctly described this as being "hit like a sledgehammer" as the rocket passes through the sound barrier.

However, a weather balloon released thirty minutes before the launch window opened found that wind shear was within the acceptable limits by a few percent. The countdown proceeded into its final seconds. With Musk on hand in the control room to supervise the launch of his new and much more powerful Falcon 9 rocket, Dunn was on the microphone.

"That was the only time I did the countdown, and I was so scared I was going to flub counting down the ten numbers," Dunn said. "I just felt so much pressure, and so much anxiety about doing the countdown. I remember having it written out in front of me and reading the numbers instead of doing it from memory. I had to make sure I wasn't going to mess that up when it was live and recorded for all time."

Dunn did not flub the countdown, although a careful listener can detect a tinge of nervousness, excitement, and fatigue in his voice. At T-minus 5 seconds the water deluge system came on, and at T-0 the rocket's nine Merlin engines roared to life. Seconds later, Dunn called, "Liftoff!" as the white Falcon 9 began its climb toward outer space against a beautiful blue backdrop of ocean and sky.

As Kellie watched from the edge of the blast zone, the rocket's sound and fury ripped across his chest. Almost immediately, and long before it was clear the launch would be successful, Chai started whooping and hollering, jumping up and down and pumping his fists into the air. Then the technician turned to Kellie and grabbed his shoulders in a half hug. Beyond exhausted, Kellie felt himself being pulled up and down, as Chai joyously shook him as if he were a rag doll.

"What the hell are you so happy about?" Kellie asked, wondering about Chai's premature celebration.

"It's off the pad," Chai shouted back. "That bitch ain't ever coming back."

And it was true, Kellie realized. The next morning, at 1 AM, he would not have to drag himself to work on the launch site, running a seemingly unending marathon to get a rocket off the pad. He started jumping up and down, too, losing himself in a moment of pure relief and release.

## The grind wears down an essential cog

In the final hour before the launch window opened, Buzza worried the countdown might need to enter a hold in the final minutes. This was not due to the upper-level winds or some technical concerns. Rather, Musk was running late. Finally, he and his then wife, actress Talulah Riley, arrived fashionably late at the launch control center about fifteen minutes before 9 AM local time. As the rocket took off, the couple went running outside to witness the fiery show firsthand.

In the immediate aftermath of the launch, it was all hugs and smiles in the control room. Musk celebrated these triumphs with his team. Yet despite the happy ending to the first launch from Vandenberg, Buzza still felt determined to step away from SpaceX after a long career.

His relationship with Musk was fraying, and the old guard were stepping away. Musk had founded SpaceX in 2002 with two employees, Chris Thompson and Tom Mueller. Thompson had left in May 2012, and by 2013 Mueller wanted to step back from running a growing propulsion department to focus on technical development. Hans Koenigsmann and Gwynne Shotwell, other critical early employees, would stay longer.

They were Musk's earliest employees, so essential in pushing and pulling the company through its desperate first days. But after finally tasting success, they found that there never came a chance to bask in that success, or catch their breath. Musk's pace and demands were relentless.

"Many of us thought after Falcon 1, and getting missions to the International Space Station, there might be something like taking our foot off the gas, at least for a little while," Buzza said. "But it didn't happen. It just accelerated."

Buzza wanted to stop the ride and get off, but he had committed to leading the first launch of the Falcon 9 version 1.1 from Space Launch Complex 40 at Cape Canaveral. This mission, for Luxembourg-based telecommunications company SES, was extra special because it marked the first time the Falcon 9 put a payload into a geostationary transfer orbit. This allowed SpaceX to start tapping into the lucrative launch market for geostationary satellites, which fly about 22,000 miles above Earth.

After suffering two different scrubs over the course of a week, the SES-8 mission launched on the evening of December 3, 2013. When the satellite successfully reached its transfer orbit, Musk came over and congratulated Buzza. He also asked the launch director to be in his office the next day.

At that meeting, Musk said he could tell that Buzza was not feeling well. And indeed, Buzza was not. He had persistent abdominal pain and nausea. Musk told Buzza he wanted to help. "I will do anything to help you get healthy," Musk said. He told Buzza to visit a doctor he knew at Cedars-Sinai Medical Center in Los Angeles. There, Buzza was diagnosed with small intestinal bacterial overgrowth, or SIBO. Stress worsens illness, and certainly there had been plenty of stress in California and Florida. The doctor prescribed an experimental drug that cleared up Buzza's symptoms. Musk understood that he had pushed Buzza hard, and he gave him three months to recover.

"He saw that I needed to get help, and he helped me," Buzza said. "That's the side of Elon I don't think everyone sees."

After three months Buzza still wanted to leave SpaceX. He could see the new guard, with new energy, stepping up to take the reins in key areas of the company. Second generation hires like Bulent Altan were leading avionics, Mark Juncosa had structures, and Zach Dunn would take over propulsion. The business was in a good spot, Buzza reasoned. He'd helped deliver the new, more powerful Falcon 9 with a splash, Dragons and commercial satellites were flying into orbit, and SpaceX had two operational launch pads on opposite sides of the country. The core of the business was ready to go. Buzza could see that the next major milestones for SpaceX, including landing and reusing Falcons, achieving human spaceflight, and creating the Falcon Heavy, were several years down the line. As glorious as those moments would be, he had a family to reconnect with. And so he did. Tim Buzza, an essential cog at SpaceX from its earliest days, stepped aside.

As ever, the Musk machine rolled onward.

|7|

# GET THE BARGE READY

*April 20, 2014*
Hawthorne, California

Zach Dunn awoke with a start.

Groggily, he recognized the tone of his mobile phone buzzing, and he groped for it in the dark. When he answered, a booming voice greeted him with a searing command:

"All J-3s remain silent! All J-3s remain silent!"

This baffled Dunn. It was 4 AM in California, on Easter Sunday. What were J-3s? Why were they being told to remain silent? Why had this disembodied voice called him in the middle of the night? The voice did not wait for Dunn's bleary thoughts to clear.

"All J-3s remain silent for roll call!" the voice insisted. And then a roll call began, with the voice calling, "Navy!" and someone replying, "Navy here!" So it went down the line, with the Army, Air Force, Coast Guard, and NORAD all affirming their presence, until it came to "SEC

DEF Rep." A representative for the Secretary of Defense indicated that he, too, was on the line.

As this roll call proceeded, the sleep fog inside Zach Dunn's head slowly dissipated. Two days earlier, on April 18, 2014, the Falcon 9 rocket had launched for the ninth time, flying a third cargo supply mission for NASA. After the launch, the rocket's first stage made a remarkably controlled descent into the Atlantic Ocean, a few hundred miles off the Florida coast.

Video from the rocket, relayed into SpaceX's flight control room by a tracking aircraft, had been tantalizing. A camera onboard the first stage showed pixelated images of the rocket as it slowed down near the ocean and softly touched down on the surface. The video had given out shortly after touchdown, and the data stream stopped altogether after eight seconds, leaving the fate of the rocket unknown. What had happened? Had the Falcon 9 sunk straight down into the depths? Had it broken into pieces? Or had it gently tipped over and now waited patiently, bobbing in the ocean, for SpaceX to come and fish it out?

There was one person absolutely dying to find out. The successful soft landing had thrilled Musk, and now he was incredibly eager to discover the rocket's fate. Almost maniacally, he began to push to reach the rocket. If it could be recovered, the first stage would provide valuable clues about how to improve future landing attempts. Moreover, SpaceX could begin to understand how much refurbishment would be needed between flights of used rockets.

"With the weather conditions and sea state I thought for sure it was destroyed," Dunn said. "But Elon held out hope, and he made it crystal clear that we were going to try to find it. He was unmovable."

Musk put on a full court press. On the Saturday morning after the launch, he huddled in his cubicle with Gwynne Shotwell and Dunn, then the director of launch and mission operations. The company's pilots had already been dispatched to the area of the ocean where the rocket had landed, but their search was complicated by a major storm in the Atlantic Ocean. They couldn't find anything.

Dunn, Musk, and Shotwell made hundreds of calls, reaching out to anyone who might be able to pull strings in the U.S. Navy or Coast Guard. They wanted to mobilize several of the Navy's P-3 Orion surveillance aircraft, which had a synthetic aperture radar that could see through the clouds and pinpoint debris on the surface of the water. Most of the calls led nowhere, but toward the end of the day Musk reached someone in the upper echelons of the Navy. Late Saturday night, with the promise of help uncertain, Dunn had returned to the corporate apartment he was staying in near SpaceX's headquarters in Hawthorne. There, his phone had shaken him from an uneasy slumber a few hours later.

After the voice confirmed the Secretary of Defense's presence, it followed up: "SpaceX Rep?"

"I vividly recall hearing myself sounding like Mickey Mouse squeaking out 'huuh . . . huhhhere . . .'" Dunn said.

To this, the voice responded, "SpaceX Rep present situational briefing." The situation was that Dunn sat disheveled, on a rented bed, in his underwear. The bedroom was dark as he had not yet had a chance to turn the lights on. Sixty seconds earlier he had been fast asleep. Now he was on the phone with a bunch of J-3s—one-star generals and other officers in the Joint Chiefs of Staff's office—as well as senior officers in the major military branches.

Dunn went through the basics of the rocket launch, its landing, and SpaceX's desire to find the booster intact, or at least what was left of it.

After Dunn's impromptu briefing, the Orion P-3s were cleared for the mission, a hard-won victory. However, just as the planes were scrambling to take off that morning, a SpaceX pilot managed to find a debris field associated with the Falcon 9 rocket. It was not bobbing in the ocean, but rather had broken apart. Regrettably, Dunn called off the P-3 flight.

This capped an intense forty-eight-hour period for Musk, Dunn, Shotwell, and others involved in the great Easter rocket hunt. These were pivotal months for the company as it moved pell-mell toward recovering the first stage of a rocket by landing it back on the planet under its own power. Nothing like that had ever been done before in the history of rocketry. Musk was convinced it could be done and wanted badly to find this rocket as proof. Some of his engineers were not as certain, but they went all in on recovery because of Musk's ironclad commitment.

"Those sorts of moments were always so tough, but also the ones in which you really learn what you are capable of doing," Dunn said of scrambling to locate the fallen rocket. "It was a classic Elon moment. We ended up not getting the stage of course, but I learned about an even deeper gear when it comes to not giving up until it is over."

## The rocky road to reusablity

Before SpaceX came along, others had tried to launch and land rockets, too, but they had given up. The U.S. Department of Defense, NASA, and America's preeminent aerospace companies had all taken their shots at developing rapidly reusable rockets. And after missing the mark, they all went back to flying expendable boosters.

NASA tried the longest. The space agency realized the importance of lower-cost spaceflight as far back as the late 1960s, when the final three Apollo missions were canceled. Even as astronauts took humanity's first steps on the Moon, the U.S. government got sticker shock from the massive Saturn V rocket, which cost billions of dollars per launch in present-day dollars. For a more sustainable future, NASA looked to the space shuttle, which the agency's human spaceflight chief said would slash the cost of sending a pound of payload to orbit to $25. Average Americans, he said, would be able to buy a ticket and go to space. "We can open up a whole new era of space exploration," George Mueller said at NASA's Space Shuttle Symposium in 1969.

With the shuttle NASA succeeded in building the world's first reusable spacecraft. The famous orbiters landed on a long runway and flew dozens of times. The tall, white boosters attached to the side of the vehicle were fished out of the Atlantic Ocean and refurbished. Only the large, orange external fuel tank was expended during launch. Although it was a technical wonder at its debut in 1981, the space shuttle fell far short of its lofty goals. The orbiters required extensive maintenance between flights. As but one example, each of the 21,000 delicate tiles so effective at deflecting heat during the orbiter's fiery reentry into Earth's atmosphere had to be meticulously inspected following each landing.

All of this work required a standing army of thousands of technicians and engineers. Two fatal accidents over the course of 135 flights, *Challenger* in 1986 and *Columbia* in 2003, only magnified the paperwork and procedures, further adding to costs. By the shuttle program's end in 2011, when its costs were tallied up, the pound-to-orbit cost was nearly $25,000—a thousand times greater than NASA's optimistic projections. Although the space agency largely met the requirement of making a reusable rocket, it had failed at the other key steps needed to make spaceflight sustainable—rapid and low-cost reuse.

The next major attempt to tackle the problem of prohibitive launch costs came in 1991, with the U.S. Department of Defense's "Star Wars" program, formally known as the Strategic Defense Initiative. A small group of proponents, including science fiction author Jerry Pournelle, convinced Vice President Dan Quayle that a space-based defense system could be serviced by a reusable spacecraft. Eventually, the Air Force funded work by McDonnell Douglas to build a one-third scale prototype of a vehicle intended to launch from the surface of the Earth, reach orbit, and land vertically. This became known as the experimental "Delta Clipper" project, or DC-X.

A small team of engineers worked quickly to develop the vehicle, which was powered by four RL-10 engines. A team of just one hundred people built the original DC-X in a mere twenty-one months, completing the vehicle by the summer of 1993. Painted white and standing thirty-nine feet tall, the DC-X bore resemblance to the bald, conical heads popularized in the Coneheads skit on *Saturday Night Live*. In August, as the rocket sat on a concrete pad at White Sands Missile Range, seven people crammed into a nearby trailer. Among them was Pete Conrad, the former Apollo astronaut who worked for McDonnell. Under their supervision, the vehicle ascended to 150 feet, moved laterally, and landed down range 350 feet. As the dust cleared, the pyramidal DC-X stood proud, its white paint covered in soot. The flight lasted fifty-nine seconds.

But after two more flights, the program lost its backing from the Air Force in the fall of 1993. As the Cold War faded, the Clinton White House ended funding for "Star Wars" initiatives. NASA stepped in to fill the gap and eventually tapped a young engineer from Purdue University, Dan Dumbacher, to lead the program.

Mindful of the need for rapidity and low costs, he sought to push the turnaround time between DC-X flights down to just eight hours, with just fifty people working on the vehicle between flights. The NASA-led team modified the vehicle to have lighter fuel tanks, targeting early July 1996 for the rapid-turnaround demonstration.

With more than 400 visitors and VIPs on-site, including NASA Administrator Dan Goldin, the first flight lifted off shortly before noon in New Mexico on June 7. The vehicle reached 2,000 feet before landing in the desert. The teams recovered the rocket and were prepping for a second flight when thunderstorms came rolling across the San

Andres Mountains, shutting them down for the day. The Army had committed the air space above White Sands to another customer the next morning, but early in the afternoon on June 8 the Delta Clipper took flight again, just twenty-six hours later, going much higher. Most of the visitors were gone, but the triumph of flying two miles into the sky was no less for it.

This feeling of elation evaporated several weeks later when, during its very next flight, one of the Delta Clipper's landing struts failed to deploy. The vehicle's liquid oxygen tank exploded on impact, and the Delta Clipper would fly no more.

The program became a casualty of rocket politics. Key officials at NASA, who had worked much of their careers on the winged space shuttle, never really liked the DC-X approach to vertically taking off and vertically landing. In a competition for further funding, a McDonnell Douglas proposal to scale up the DC-X lost to Lockheed's winged X-33 concept for a space plane.

"The agency went to the X-33 because we had great hopes for a single-stage-to-orbit space plane that landed like an airplane," Dumbacher said. "It was the end of the road for vertical landing technology."

The X-33 was not bound for a happy fate either. The technological challenges were many, and in particular engineers had a difficult time building fuel tanks to store liquid hydrogen at extremely cold temperatures. Although NASA had spent nearly $1 billion by the year 2000 on the program, much work remained before any flights could be attempted. Its future was further muddied as NASA initiated yet another program, called the Space Launch Initiative, to develop technologies for reusable rockets. This meant that, for the 2001 budget cycle, NASA sought money for three separate reusable launch projects, the space shuttle, the X-33, and an entirely new launch system. The White House Office of Management and Budget threw a fit.

The party would soon end. NASA canceled the X-33 in February 2001. Two years later, space shuttle *Columbia* disintegrated as it glided back to Earth following a sixteen-day mission in orbit, killing all seven astronauts aboard. After this second disaster, the space shuttle's days were numbered. Then, in 2004, NASA abandoned the Space Launch Initiative. Only a few years earlier NASA had believed so deeply in low-cost, reusable spaceflight that it had funded three separate programs to ensure its future. Now, the technology had no future at the world's pre-eminent space agency. NASA was going back to the Apollo era of big, expendable rockets.

Musk founded SpaceX in 2002, as NASA abandoned attempts to develop low-cost, reusable spaceflight. He saw an aerospace industry largely unwilling to pursue innovative ideas without the guarantee of government funding. McDonnell Douglas could have continued the Delta Clipper on its own. Lockheed could have pushed ahead on the X-33. Had either company got their vehicle flying, NASA and the military would have paid handsomely for their services. But as traditional contractors, these companies were accustomed to fat government contracts up front.

For a time, that strategy worked. For nearly two decades longer McDonnell Douglas, which was acquired by Boeing, and Lockheed would continue to feast on the government's largesse when it came to space contracts. They would do so right up until SpaceX and its low-cost, reusable Falcon 9 rocket ate their lunch.

## Fireballs, grasshoppers, and Johnny Cash

Recovery and reuse were part of the SpaceX approach from the beginning. The very first Falcon 1 rocket launched with a parachute stuffed into the interstage. Musk even dispatched an engineer onto an Army boat for recovery of the first stage. This proved a forlorn task, as the Falcon 1 was a tiny white cylinder smaller than a city bus, plummeting into a sea of white caps. It was impossible to find such a needle in the haystack.

Since the Falcon 9 rocket was much larger, Musk reasoned that it should be easier to observe coming back from space. So in June 2010, he sent his Dassault Falcon jet to monitor the rocket's reentry and guide a recovery vessel in the ocean. Onboard was Abhi Tripathi, who had joined SpaceX just three weeks earlier to assist with Dragon operations and help it meet NASA's requirements.

After a launch, the first and second stages of the Falcon 9 separate near the edge of space, at an altitude of about fifty miles. The first stage continues rising for a time, under its own momentum, before succumbing to gravity and falling back to Earth. Not a whole lot of planning went into the first Falcon 9 recovery attempt, so the pilot flew toward the region cited in the mission's Notices to Airmen, or NOTAMs, which outline potentially hazardous areas that aircraft and boats must avoid during a rocket launch. For the first Falcon 9 launch, much of this area lay over the Bermuda Triangle.

Tripathi and two pilots listened to the cockpit radio as the rocket took off. Excitedly, they descended to about 500 feet. As they circled above the choppy seas, Tripathi's mind drifted to urban legends about all of the aircraft and boats lost in this area of water south of Bermuda. The aircraft's flight computer seemed unsettled, too. In its synthesized voice the computer kept repeating, "Pull up, pull up, pull up." As the little plane flew larger and larger circles, Tripathi realized his search was in vain.

"My sole job was to take these binoculars and look for rocket debris," he said. "I did not see one damn piece of debris. And even if there was debris, I don't think you could have differentiated it from the whitecaps. The rocket must have shattered so bad there was nothing left."

Most SpaceX engineers viewed recovering a rocket that was coming back from that altitude, equipped with but a single parachute, traveling at many times the speed of sound, as a hopeless task. Atmospheric forces would batter the rocket to smithereens before its parachute had a chance to deploy. But Musk believed there was a chance that large chunks, at least, would land in the ocean and could be salvaged. Even if there was a 10 percent chance of survival, he said, SpaceX should attempt to recover the rocket.

A young aerodynamics engineer named Justin Richeson was tasked with reviewing data from the first Falcon 9 launch and improving the odds of surviving reentry during the second flight. Richeson's team identified a number of problems, the biggest of which was atmospheric heating. For the second Falcon 9 launch, therefore, SpaceX added cork insulation around much of the first stage, including the engine section, hoping this material would take the brunt of heating as the first stage fell back to Earth.

Richeson joined Tripathi as a debris spotter for the Falcon 9's second flight. This time, with more advance notice, Tripathi did his homework. From the rocket's planned trajectory, he calculated precisely where the Dassault jet should be to observe reentry. When the rocket launched on the morning of December 8, 2010, Tripathi was well positioned.

This time, he did not miss it. Peering intently through the jet's windows, Tripathi looked to where he believed the first stage would likely come back down from space. And then he saw it—a fireball streaking across the sky. SpaceX's first stage created quite a show as it burned up, broke apart, and plummeted toward the Atlantic Ocean. About five minutes later the satellite phone in the seat next to Tripathi began to ring. It was Musk calling.

"Did you have eyes on it?" Musk asked.

Tripathi had, and he told the boss what he had seen.

"Shit," Musk replied. "Okay, thank you." He hung up.

After these early recovery failures, Musk understood that he needed to try something other than parachutes. He found inspiration from what other new space companies were working on at the time. In 2009, NASA had funded a Lunar Lander Challenge to see if the commercial space industry could launch a small rocket, ascend to about 150 feet, and land it at a nearby pad. The vehicle then had to be refueled and fly back to its original launch pad within a couple of hours.

A small, California-based company called Masten Space Systems won the competition in late October 2009 with a small rocket, taking home $1 million. Armadillo Aerospace, from Texas, took second place. Musk was good friends with the founder of Armadillo, video game developer John Carmack, so he was aware of the competition.

Another important milestone came in 2010, when Masten built and flew its Xombie vehicle at the Mojave Air and Space Port in California. After launching a few hundred feet into the air, the small vehicle shut down its engine and then relit it a few seconds later to enable a soft landing back from where it took off. No rocket had ever taken off vertically and performed an in-air relight of its engine. This took NASA's work with the DC-X rocket a decade and a half earlier to the next level. That Masten had pulled this off with such a small team, and almost no budget, showed that SpaceX might be able to self-fund propulsive landing rather than needing a major government project like DC-X.

"We were a small company, really just five guys and a few interns, and we came out of nowhere to beat John Carmack in the Lunar Landing Challenge," said Jonathan Goff, a cofounder of Masten. "Then we did an in-air relight. I think that's what woke Elon up to the fact that he didn't have to spend a fortune on this."

Following two failures with a parachute on the Falcon 9, Musk called a weekend meeting in 2011 with several key engineers to strategize options for getting the Falcon 9 first stage back in one piece. They settled on a vertical landing of the rocket, under its own power. This approach had many downsides. Any fuel used to slow the rocket down meant less propellant would be available for launch. Adding landing legs and other structures to support a vertical landing would also impose a mass penalty. Economically, it may not make sense to whack that much payload capacity in the name of reuse.

As SpaceX was still in the infancy of launching the Falcon 9, testing landing legs on an operational mission seemed too risky. Musk and the engineers also had serious questions about firing a Merlin engine close to the ground, such as whether its altitude sensor could make accurate readings through dust and exhaust. They needed some kind of a testbed. At the weekend meeting, Musk asked Chris Hansen, an engineer who led testing of structures, to make one.

"My team and I had responsibility for building a variety of test stands, so he asked me to go and build a flying test stand," Hansen said.

He led a small team. Doug Bernauer put together the avionics. Shana Diez, a propulsion engineer, did a lot of the plumbing to flow propellant from the tank to the single engine. They scrounged up parts in McGregor, using the old Falcon 1.0 first stage qualification tank and a Merlin engine that had undergone a spate of development tests. The avionics components were actually pulled out of the mothballed C1 Dragon spacecraft hanging in Hawthorne. The resulting clunky, kludged-together vehicle stood about 100 feet tall. At first they named the project "Hopper." Then Hansen started calling the vehicle "Grasshopper" because they were testing it in a field in Texas. The name stuck.

Musk initially wanted Grasshopper ready to fly in six months, but it took a full year before the team was ready to start flying in September 2012. Everyone had so much else to do. The Grasshopper program strained a workforce already pushed to its limits by development of version 1.1 of the Falcon 9 rocket and trying to get Cargo Dragon to the space station. The propulsion, avionics, and software departments had no slack.

"Nobody within SpaceX, except for Elon and a few other people, thought Grasshopper was a good idea," said Robert Rose, the director of flight software. "People treated it like this giant waste of time. SpaceX was strapped for cash, and we didn't have enough resources to do all of the things we wanted to do. Elon was on our case night and day about increasing launch cadence and getting Dragon flying."

Engineers openly debated the merits of reuse in the halls at SpaceX and grumbled about Grasshopper. In addition to the mass penalty for extra fuel and landing legs, it was not clear whether rockets could be refurbished easily, because it had not worked with the space shuttle. And would customers want to fly on a reused rocket?

Grasshopper's landing legs being worked on. | PHOTO CREDIT: CHRIS HANSEN

The early Grasshopper tests were nevertheless crowd pleasers back in Hawthorne. Employees gathered on the third floor to watch on a big-screen projector. The festivities were enhanced when two avionics engineers, Edwin Chiu and Bryan White, broke out a violin and a guitar, respectively. As Grasshopper climbed and then descended, they theatrically sang Johnny Cash's "Ring of Fire," loudly emphasizing the chorus:

*I fell into a burning ring of fire*
*I went down, down, down*
*And the flames went higher*
*And it burns, burns, burns*

During the early flights in the fall of 2012, some SpaceXers murmured and cheered when it looked as though Grasshopper might crash. There was some hope that, if the vehicle proved a disaster, they would not have to work on the project any longer. By the time of these first flights, to focus on test stands and ensure he was in California for the birth of a daughter, Hansen passed control of the program to Diez. They timed the handover to when Diez returned from a two-week vacation. When they discussed this transition timing with Musk, he could not believe anyone would want to take a vacation for that long. Too boring, he thought.

Having come to SpaceX from Blue Origin, Shana Diez became chief engineer of the test flight program and brilliantly guided Grasshopper higher and higher. During its final flight in October 2013 the vehicle ascended to 2,400 feet before safely touching down. By then the cheers in Hawthorne for her and the Grasshopper team were genuine. The scruffy test vehicle had taught SpaceX much, including the effects of firing an engine close to the ground, measuring altitude through dust and debris, rapid throttling of engine thrust, testing out landing legs, and landing with partially full fuel tanks. This would inform not just the Falcon 9 program, but Starship, Musk's vison for traveling to Mars, as well. It is no coincidence that Diez ended up leading Starship engineering.

During these years, in company-wide gatherings, Musk would deliver his usual spiel about founding SpaceX to make life multiplanetary, to settle Mars, and so on. Justin Richeson, who was helping to lead reentry efforts, nodded along with the rest of the company's employees. "It felt like it was kind of a joke," Richeson said. "We were like, 'Yeah, whatever.'" But the success of Grasshopper and Musk's unwavering conviction eventually won people over, including Richeson.

Musk attended weekly meetings of the reentry team in his executive conference room to receive updates on progress. At the end of one in 2013, Musk got up to leave. But as he walked toward the door he stopped, turned around, and looked intently back at his engineers. "It may seem like these are small steps," he said. "But we are not going to Mars in my lifetime, or yours, if we don't get our act together and take this first step."

"It was intense," Richeson said. "It was the most sincere I had ever seen the man. That was the first time I understood this dude was not kidding around, that this was literally the entire mission for the company."

Musk took another key step toward resolving the vertical landing challenge when he hired an engineer named Lars Blackmore from NASA's Jet Propulsion Laboratory in 2011. Blackmore had co-invented an algorithm—essentially a set of rules and detailed instructions—to guide a spacecraft's powered descent down to a planetary surface, an algorithm that was both accurate in terms of location and minimized the amount of propellant consumed. Although this G-FOLD algorithm applied primarily to setting rovers down on Mars, the same principle could be adapted for tall, skinny rockets returning through Earth's atmosphere.

At the moment it separates from the second stage, a Falcon 9 stage 1 rocket is happily soaring above the atmosphere at Mach 9, which is more than ten times faster than a commercial airliner. Before attempting a landing, it must somehow survive the extreme heat of passing back through Earth's atmosphere without getting battered to pieces in the process. Crumpling is a major concern. Relative to its overall size, the rocket's aluminum structure—its outer, protective skin—is thinner than a beer can. As it comes back through the atmosphere, the rocket must hold a precise angle of attack or sideways pressure will crush it.

Musk's team found that steering a rocket through the upper atmosphere proved difficult. In the thin air, when not actively under thrust from Merlin engines, the first stage easily lost control. To address this, engineers added small thrusters powered by compressed nitrogen near the top of the first stage. As the rocket descended, these "cold gas" thrusters would periodically emit small puffs of nitrogen to properly orient the booster. This helped, but the simple thrusters did not have enough oomph. What sealed the deal was the addition of four "grid fins," each measuring four-by-five feet and looking something like a flat cheese grater, that could roll, pitch, and yaw up to 20 degrees. This provided the precision steering needed to hit a landing site.

Another major problem fell to Mueller and the propulsion team. During its return, the rocket's engines needed to perform three separate burns in a variety of environments. Mueller had experienced so many setbacks trying to light engines on the ground, and now they had to relight the engines three times: in space, at the edge of the upper atmosphere, and in the lower atmosphere. Initially, after stage separation, a "boostback" burn of one or three Merlin engines would slow the rocket and put it on a trajectory to either return to land or alight on the deck of a drone ship offshore. Then, at the top of the atmosphere, an "entry" burn of three engines was needed to slow down the rocket. Finally, near the ground, a single Merlin engine in the center would relight a final time to bring the rocket nearly to a hover before touchdown.

Previously, NASA had modeled the ignition of engines in these environments through computer simulations, giving it the fancy name "supersonic retropropulsion." But in the real world, no one had ignited engines while flying backward through the atmosphere at supersonic speed. Can you imagine that? A rocket is falling through the sky at incredibly high speeds, air screaming all around, and temperatures rising above 1,000 degrees Fahrenheit (538 degrees Celsius). Now, light the engines—preferably without blowing the rocket up. Mueller and his team had to turn scientific theory into real-world practice.

The blunt end of the rocket, which becomes the forward end as it returns to Earth, was a serious impediment. The fastest airplane ever built, Lockheed's SR-71 "Blackbird," had an aerodynamic design with a long, sleek nose. The back side of the Falcon 9 rocket, with its nine large engine nozzles sticking out, seemed hopeless.

"You've got nine Dixie cups as your aero surfaces," Mueller said. "It's the most un-aerodynamic thing you can imagine, going three times as fast as the SR-71, which looks like a needle. The loads are tremendous, and the outer engines just wanted to get splayed out. So after the entry burn, the outer engines would tuck in like tail feathers, and be locked, to get as compact as possible."

The company's first real-world test of all this came during the CASSIOPE mission, the trouble-prone initial Vandenberg launch, in September 2013. As part of the move to version 1.1 of the Falcon 9, the propulsion engineers had not just been tasked with upgrading to the Merlin 1D engine by making it more efficient and more cost effective. Mueller's team also had to make sure it survived reentry, as well as multiple ignitions at high velocity.

Mueller went to the launch site to watch the mission alongside Buzza and Koenigsmann, the core of the company's original, brilliant engineering team. Under Musk's supervision they had led departments, built rockets, and, for the most part, safely launched them for more than a decade. Now they wanted to see if landing and reusing those rockets might be possible.

After the successful launch, the trio tracked a video feed from Musk's plane that observed reentry. This time, there was no fireball. Instead, the rocket successfully fired three engines to slow its descent through the atmosphere. While the landing burn did not occur entirely as intended, the rocket still made it down to the surface of the ocean in one piece before corkscrewing and slamming into the water.

In that single moment Mueller, Buzza, and Koenigsmann understood that reuse was not just possible, but that they were closer to pulling it off than anyone realized. Musk had been planning to make much higher Grasshopper-like flights from White Sands Missile Range in New Mexico. In McGregor they had been confined to an altitude up to about half a mile. In New Mexico they could go above sixty miles. The new site was three-quarters built by the time of the CASSIOPE mission. But it was never completed. After the Falcon 9 hit the ocean, SpaceX no longer needed such testing. Musk decided that refinements to the landing process could take place on real rockets coming back from space.

"I remember watching the live video and seeing the light of the engine on the ocean," Mueller said. "And holy shit, it was there. The rocket came down, landed in the ocean, and blew up. That was unreal. It worked the first time. I was like, get the barge ready. Get the landing legs ready. This shit works."

## Just Read the Instructions

It would fall to Ben Kellie, among others, to get the barge ready.

He had put everything he had into the pad build at Vandenberg. The work had been breakneck, but meaningful. After coming off the adrenaline high of the CASSIOPE launch, however, there was a void he could not fill. As the company's launch focus turned to Florida, he realized the next mission from Vandenberg was years away.

"Leading up to the launch we worked like feral psychopaths," Kellie said. "Then, afterward, they wanted us to just start writing procedures for the Air Force. We sat in meetings all day. That's not what we were good at."

Returning to "normal" life after a major event at SpaceX was a huge challenge. During months of intense activity, everything an employee did mattered, and the livelihoods, and sometimes the lives, of coworkers were on the line. One had to be tenacious, smart, and creative, and ready to make decisions in a split second. The responsibility was both exhilarating and exhausting. And then, in a matter of minutes after a launch, life suddenly slowed back down, without transition. The night after a successful launch was the highest high. The days after were very deep troughs. SpaceX was not a job; it was a lifestyle.

Kellie's friends started leaving Vandenberg. Zach Dunn and Lee Rosen moved down to headquarters in Hawthorne. It was no longer much fun without the *Dance Dance Revolution* crew. Neither was Kellie's personal life. His marriage was on the rocks, and although he knew it probably was too late, he kept a vow to his wife to quit the crazy job and its insane hours. Two years had felt like two decades. He departed SpaceX in July 2014. Kellie moved to Seattle and struggled with the transition. His marriage failed.

Therefore, when a former colleague from SpaceX called two months later, he listened. The offer of a fresh, new challenge was tempting. SpaceX had rented a barge in a small, sleepy town in southern Louisiana. Someone needed to go down to Morgan City and push like hell to retrofit the ship for rocket landings. He had eight weeks to prepare the barge and move it to Florida. Kellie packed a single suitcase and caught a flight to Louisiana the next day.

When a hurricane makes landfall, Morgan City is one of those towns most vulnerable to being submerged. Built along the banks of the muddy Atchafalaya River, it sits just six feet above the Gulf of Mexico. But due to its proximity to the water, Morgan City has large shipyards. SpaceX had leased the Marmac 300 barge from a tugboat company called McDonough Marine Service. As the name suggested, the barge was 300 feet long and 100 feet wide. It was not the best barge in McDonough's fleet. Already more than fifteen years old, it had been pretty well worked over. More colorfully, Kellie characterized it as "beat to shit." Unfortunately for the Marmac 300, SpaceX would beat it to shit even more. By crashing rockets into it.

When Kellie drove into Morgan City, he found that nothing had yet been built onto the Marmac 300 for rocket landings. In typical SpaceX fashion, he had to improvise on the fly. For example, after the rocket touched down, a compressor would need to pump air into the vehicle to purge dangerous gases from its tanks and engines, thereby stabilizing it so people could approach without risk of leaky fuel or igniter fluid fires. Before accepting the assignment, Kellie had been told all the engineering work for such systems was completed. And true enough, there was a design for this compressor system on SpaceX letterhead. However, the schematic consisted of a circle labeled "air compressor," with a line to a square. Inside the square, "TBD"—to be determined—had been written.

Such are the days of SpaceX engineers' lives.

In his first week, large "wings" arrived, which had to be welded onto the side of the barge to extend its width for a rocket landing. Later, the guidance, navigation, and control team would ask for still more area—another ten feet on each side. The total width grew to 150 feet. By early November, most of the basic work had been completed, with the installation of the wings and other subsystems needed to support a rocket landing. So the *Elsbeth III* tug boat pushed the Marmac 300 out into the Gulf of Mexico, and ten days later, on November 17, they arrived in Jacksonville, Florida.

SpaceX wanted to stage its recovery operations out of Port Canaveral, just a few miles away from the company's launch site in Florida. But the port had asked for too much money for fees and did not really take the rocket company seriously. So SpaceX shifted recovery operations 150 miles up the coast to the Port of Jacksonville. They tied the barge up right next to the cruise ship docks. Every seven days a Disney cruise ship would pull up to the barge, nose to nose. Even in November it could get hot on the barge deck in Florida, when pulling cable and performing myriad other tasks. Kellie worked at a picnic table set up in the middle of the deck, beneath an umbrella, typing on a laptop. When the wind was right, he could hear conversations from passengers on the Disney ship, high above, looking down on the SpaceX proletariats slaving away in the heat. When Kellie overheard a parent explaining to his son that he needed to go to college so he would not have to work like that, Kellie turned around and shouted back, "Hey, I went to college twice!"

The attitudes of onlookers started to change when the purpose of the barge became clear. Initially, Musk had not wanted to put a large X in the middle of the barge, as in "SpaceX" and "X marks the spot." Musk believed the company would look stupid when rockets started smashing into what looked like a target. After considering alternatives, such as an "F" for Falcon 9 or the Falcon 9 logo, Musk relented. "Do the X," he declared in a classic, one-line email to Kellie. He also supplied a name for the barge: *Just Read the Instructions*, after the sentient starship in the Culture series of science fiction novels by Iain M. Banks.

As 2014 ended, it became clear the drone ship would soon be called into service. After fits and starts with landing attempts, SpaceX engineers began to have enough confidence to try hitting the "X" on a drone ship. With its fourteenth overall Falcon 9 launch, a Dragon cargo delivery mission planned for early January, SpaceX therefore would try landing at sea. Kellie believed *Just Read the Instructions* was about as ready as it was going to be for handling dive-bombing rockets.

A day before putting to sea for its first recovery mission, Kellie was showing a fellow SpaceX employee, Robb Kulin, around the vessel. Kulin still worked for Mark Juncosa and had come to Jacksonville to see what needed to be done to push the landing capability over the line. During the visit, Kellie asked what the company's plan was if the rocket actually landed. There were so many details: securing the rocket, lifting it from the barge to the docks, preparing it for transport to Cape Canaveral, and more. Kulin explained that some of these details remained in flux.

Perhaps the most immediate concern was finding a place to offload the rocket from the barge.

This response stunned Kellie. SpaceX was possibly mere hours from landing a rocket. He texted Dunn, asking about pedestals on which to place the recovered first stage at the docks. A few minutes later he received a forwarded email, a long chain of responses with about forty SpaceX engineers arguing about where to put the pedestals. The Falcon 9 rocket's first stage had four reinforced pins at its base, where the rocket is clamped down at the launch site. These were hard points on the vehicle. Four large pedestals, essentially blocks of steel, had to be installed in the port as a base for the rocket. A land recovery team at the Cape, led by an engineer named Trip Harriss, had designed and fabricated these pedestals. But no one had ever measured locations in Jacksonville to accommodate these pedestals or the rocket. And this was no small matter, as the vehicle's size severely limited the available options.

Instead of putting out to sea with *Just Read the Instructions* the next day, Kellie and another SpaceXer, Chris Newton, stayed behind. Like Kellie, Newton was a jack-of-all-trades engineer who preferred to work in the field and could be thrown into any task and expected to take ownership of a problem. The pair promptly found a crane operator to see how much space would be needed to lift the rocket off the barge, swing it onshore, and set it on the pedestals. Then they identified where the pedestals should go and lined up contractors to start digging a hole for the base of the concrete the next morning.

That evening Kellie and Newton retired to their hotel nearby, a Courtyard by Marriott. They had already spent a dozen hours that day putting the barge to sea. Now they got out their laptops and did months of detailed design work for the pedestals in a single evening. Around midnight they switched from ordering glasses of wine to ordering bottles. Dunn checked in that night and, drunkenly, the duo assured him that the contractor would be on-site at sunrise to start prepping for the pouring of concrete. After a few hours of uneven sleep, the two engineers pushed through hangovers and worked through the day with steel, rebar, and concrete. At the same time, a convoy was hastily assembled at the Cape to bring the pedestals and a mobile command trailer north. It all came together by nightfall.

This illustrates how SpaceX flew by the seat of its pants in developing a brand-new thing the world had never seen: a rocket that, however improbably, *could land on a boat*. The company's mostly young engineers did not know what they did not know. Everything was new, challenging, and complex, down to the smallest detail. They were building things, making shit up as they went along, and completing tasks just in time. SpaceX innovated so fast that it forced its engineers and technicians to solve whatever problem was most urgent that day.

"It was brutally hard work," Kellie said. "Physically hard. Mentally hard. Emotionally hard. But it was also one of those things that if you want to go on adventures, there's not a much better place to do it."

Like a lot of his fellow engineers at SpaceX, Kellie worked more than 100 hours a week during the drone ship build-out. He had packed a single suitcase for an eight-week assignment in Louisiana, and by the time of the first landing attempt, he was already entering his fourth month back at SpaceX, still living out of that same suitcase.

There were so many who contributed to the effort of developing version 1.1 of the Falcon 9 and pushing the vehicle through those early tests and failures. It is impossible to list them all, as thousands of people worked at SpaceX. But a few names who came up in interviews were engineers like Mike Rossoni and John Lindauer, who built new turbopumps for the Merlin engine from scratch; engineers such as Darin Van Pelt, Eric Murray, and Will Heltsley in Merlin development; and avionics engineers Jon Barr, Dennis Fong, and Kenny Boronowski, who worked on thermal protection for reentry. They all believed in the importance of what they were doing.

"We realized that this made a huge fucking material difference in the world," Kellie said. "We're not just chucking shit into the ocean and waving goodbye anymore. We're reusing hardware. We're dropping the cost of access to space. This was important for humanity. So we all put a shitload of effort into it because we believed what we were doing."

## A very different kind of recovery mission

So why was SpaceX putting so much effort into landing rockets at sea instead of dry land? The common conception of a rocket launch is that a vehicle blasts off from the surface of the Earth and travels straight up into the sky. This is true, but only for a few seconds. Most launches strive to reach low-Earth orbit, where objects are zipping around the planet at 17,500 mph. The goal of a launch, therefore, is not so much to push an object a few hundred miles above the Earth, but to impart a tremendous horizontal speed. If an object does not reach orbital velocity, it will simply fall back to the planet.

Very quickly after launch, then, a rocket turns mostly sideways and travels almost horizontal to the planet's surface. And therein lies the problem. Consider all of the energy expended by the first stage of a Falcon 9 rocket during its nearly three minutes of flight to push its payload toward orbit. It's pretty mind-boggling. The first stage itself does not reach orbital velocity, but it is traveling at 4,000 mph, and it is flinging an upper stage, payload fairing, and a satellite that, combined, weighs around 120 tons. Congratulations, you just threw the world's largest javelin hundreds of miles out over the ocean.

Now go and get it back.

"I really can't emphasize horizontal velocity enough, because it's so counterintuitive," Musk said. "Basically, a rocket at stage separation is zooming out to sea at an incredible velocity. And it doesn't have enough propellant to zero out that velocity, boost back in the other direction, and land. In order to achieve effective reusability, you really need an ocean landing."

Returning rockets to the launch site would simply cost too much fuel. The answer SpaceX settled on was the automated drone ship. It seems pretty mad, but with sophisticated avionics onboard the rocket, the company's engineers believed the ship could pitch and roll as much as 5 or 6 degrees on the sea and still support a landing. The drone ship would need to hold to an absolute position at sea based on GPS data, with an accuracy of three feet or less. The task for SpaceX engineers therefore was this: launch a rocket to the edge of space traveling thousands of miles an hour, slow it down, and plop it down on a rolling target at sea about the size of a basketball court.

As Kellie and Newton prepared the pedestals in Jacksonville, *Just Read the Instructions* was towed out to sea for its first attempt to catch a falling rocket in early January 2015. The rocket made a controlled descent, but it ran out of hydraulic fluid, which was necessary to move the grid fins. This was SpaceX's fifth operational cargo mission for NASA, and it launched in the predawn hours from Florida. The video of this landing was widely enjoyed inside SpaceX at the time because it shows the first stage brilliantly lit at night, striking one side of the drone ship before sliding across the deck and culminating in a spectacular explosion.

"We had to rebuild the barge completely after that landing, which I called Leeroy Jenkins because it just went completely sideways across the deck," Kellie said. "We were doing a lot of rebuilds."

The reentry team realized they needed more hydraulic fluid on the first stage. The rocket that failed to land on January 10 had an "open-loop" circuit for hydraulic fluid, which cost less to implement but was much less efficient. Musk directed his engineers to switch to a closed-loop hydraulic system for the next flight. While technically it was possible to make this change in a few weeks, there was a major problem. The next launch on the manifest was the company's very first mission for the U.S. military, bearing a small Earth-observing satellite named DSCOVR.

The Department of Defense is an essential customer for almost every U.S. launch company, but also a demanding one. The military likes to thoroughly understand the rockets it is flying on and wants to see them launched in the same exact configuration a few times before entrusting its missions to a particular launch vehicle. And here was SpaceX, on the eve of the launch, seeking to make a substantial change to its hardware.

"Elon told us to find a way to get it done," said John Muratore, who served as launch chief engineer for the mission.

So Muratore and Gwynne Shotwell went to meet with Lieutenant General Samuel A. Greaves, commander of the Air Force's Space and Missile Systems Center, which was responsible for safely delivering satellites into space for the military. Greaves was based in El Segundo, California, not far from the SpaceX factory. Muratore and Shotwell offered him and his team full visibility into the modifications to the

SpaceX headquarters in Hawthorne, California. | PHOTO CREDIT: STEVE JURVETSON

The business end of the Falcon Heavy rocket that launched NASA's Psyche mission to the asteroid belt between Mars and Jupiter. | PHOTO CREDIT: NASA

Bulent Altan works on Falcon 9 avionics. | PHOTO CREDIT: CATRIONA CHAMBERS

Roger Carlson takes a selfie atop the lightning tower with the Falcon 9 below, before its public debut at Cape Canaveral. | PHOTO CREDIT: ROGER CARLSON

A Falcon 9 first stage is lifted onto the tripod in McGregor, Texas. | PHOTO CREDIT: CATRIONA CHAMBERS

Roger Carlson stayed up all night watching to ensure the crane-dangled engine section didn't fall. | PHOTO CREDIT: ROGER CARLSON

The Falcon 9 first stage passes beneath a wire during its long journey from Texas to Florida in 2009. | PHOTO CREDIT: ROGER CARLSON

Falcon 9 first stage strikes a building in Louisiana, November 2009. | PHOTO CREDIT: ROGER CARLSON

Falcon 9 going vertical for the first time in January 2009. | PHOTO CREDIT: ROGER CARLSON

All of the elements of the first flight-ready Falcon 9 rocket in February 2010. | PHOTO CREDIT: ROGER CARLSON

Elon Musk and Chris Thompson in the Falcon 9 flight one control room. | PHOTO CREDIT: ROGER CARLSON

Party on the Cocoa Beach pier after Falcon 9 flight one. | PHOTO CREDIT: HANS KOENIGSMANN

Tim Buzza, left, shakes hands with Brigadier General Susan Helms, commander of the 45th Space Wing, at the SLC-40 groundbreaking ceremony. | PHOTO CREDIT: TIM BUZZA

Employees watch the Falcon 9 1.1 static fire test in September 2013 at Hawthorne. | PHOTO CREDIT: STEVE JURVETSON

The first Falcon 9 is worked on at SLC-40. | PHOTO CREDIT: TIM BUZZA

After landing, a Falcon 9 sits atop the *Just Read the Instructions* drone ship. | PHOTO CREDIT: NASA

The Grasshopper test vehicle in McGregor, Texas, in October 2012. The streak in the sky to the right is the CRS-1 Dragon spacecraft. | PHOTO CREDIT: DENNIS UNDERWOOD

Robb Kulin standing in front of the landed first stage after the ORBCOMM-2 launch. | PHOTO CREDIT: HANS KOENIGSMANN

A view of the Falcon 9 control room after the ORBCOMM-2 launch in December 2015. | PHOTO CREDIT: RICKY LIM

Kathy Lueders celebrates Demo-2 docking with the International Space Station. | PHOTO CREDIT: NASA

Hans Koenigsmann and Kathy Lueders hug at Demo-2. | PHOTO CREDIT: NASA

Matt Desch, hand on chest, watches Iridium-1 launch in 2017. | PHOTO CREDIT: MATT DESCH

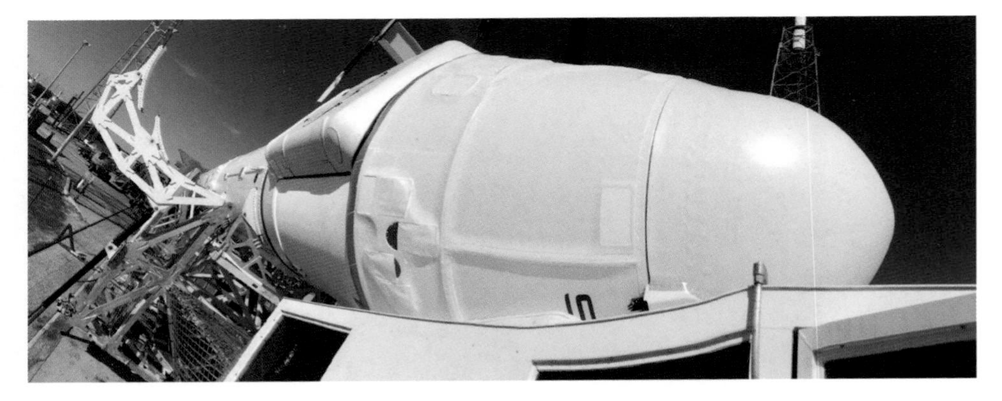

View of the CRS-3 Dragon spacecraft after being mated to the Falcon 9 rocket. | PHOTO CREDIT: HANS KOENIGSMANN

Kevin Mock leads the Dragon recovery team back to the improvised barge after its first test flight in 2010. | PHOTO CREDIT: SPACEX

The perilous crossing from the *Gladys S* to the barge. PHOTO CREDIT: ROGER CARLSON

The Crew Dragon flown in Demo-1 is recovered at sea. | PHOTO CREDIT: ROGER CARLSON

Dragon from the C1 mission hanging in the SpaceX factory. | PHOTO CREDIT: STEVE JURVETSON

The Dragon capsule used for the Crew-3 mission at Kennedy Space Center. | PHOTO CREDIT: NASA

Bob Behnken, left, and Doug Hurley inside Crew Dragon during training. | PHOTO CREDIT: NASA

SpaceX investor Steve Jurvetson, left, with Elon Musk at the McGregor Dairy Queen. | PHOTO CREDIT: STEVE JURVETSON

The Falcon 9 used for flight six, at Vandenberg Air Force Base in 2013, with the CASSIOPE satellite. | PHOTO CREDIT: HANS KOENIGSMANN

rocket. They could sit there right in the hangar and watch everything the SpaceX technicians did. Not all government officials would have said yes, but Greaves did.

"He told us that if we could reuse these rockets, it would be hugely advantageous for the Air Force, even if it meant taking a risk with the current mission," Muratore said.

The DSCOVR satellite successfully launched on February 11, a month and a day after the previous flight, with a closed-loop hydraulic system. However, stormy weather and twenty-three-foot waves precluded a drone ship landing attempt. Even so, the rocket made a vertical landing into the ocean within about thirty feet of the targeted area, indicating SpaceX had solved the hydraulic issue.

In between landing attempts, Kellie and the rest of the rocket recovery crew game-planned what they would do if they actually snagged a rocket. The strategy they settled on was, to put it mildly, slightly hazardous. The rocket would land and then dump its extra fuel directly onto the deck of the drone ship. If there was an excess of fuel, as anticipated, the deck would be sprayed with water cannons. Then the welders would come onboard and secure large eye bolts into the deck.

"It was a mess of a plan," Kellie said. "After you sprayed the deck you're in the position of having to dry the deck so you can weld it, all the while with a rocket looming tall and the barge swinging back and forth."

Should the welders succeed, the recovery team had a small Genie boom lift, and some lucky participant would climb into its small basket and be raised to a point to throw a Kevlar rope around the rocket, which would be bound through the two eye bolts on the deck. Only later on would SpaceX engineers devise large, steel "shoes" that could be welded down over the rocket's landing legs that were both more secure and safer to install.

Kellie was at sea for the next drone ship landing attempt, for the sixth NASA cargo mission, in April 2015. This one almost made it safely down, but a stuck valve caused the stage to have too much sideways motion after it reached the drone ship. As the rocket slowly slid laterally, a landing leg broke and the vehicle toppled over. This was clearly suboptimal, but one of the landing legs escaped the resulting fireball and got trapped in the handrails around the edge of the barge. Kellie and his team lashed it to the side of the barge with Kevlar rope, and it flopped in the ocean all the way back to the port. After analyzing this debris, engineers improved the system that locked the landing legs into place.

*Just Read the Instructions* returned to sea two and a half months later, in late June 2015, to try again. This would be the third drone ship landing attempt, and as Kellie and the other members of the recovery team followed along on the *GO Quest* support ship, they were eager to see what would happen. They could feel themselves getting closer to a successful landing.

Watching a rocket falling back from space, out to sea, is about as surreal as one might imagine. From the vantage point of the support ship, the first stage looked about the same size as an index finger held out at arm's length as it dropped toward the ocean. "It's impossible to believe," Kellie said. "But then you see it, and there it is. You realize that this is exactly what you've been working toward, and it's doing what it was supposed to do."

He and most of the other engineers watched from the main deck of the support ship. Kellie's primary job was monitoring the satellite antenna dish that tracked the launch and position of the rocket. The instrument came alive when a rocket took off, and it followed as it passed overhead down range. Then the dish would stop and swing back, tracking the rocket, before pointing straight up.

This meant it was time to look skyward.

The CRS-7 supply mission for NASA launched from Florida at 10:21 AM local time on June 28, 2015. The satellite dish began rising and continued doing so for nearly two and a half minutes. Then it stopped and dropped, as if dead. Kellie ran onto the bridge of the ship, asking if communications were lost with the rocket. Such "ratty" comms happened from time to time at sea. No one was sure what had happened.

Moments later the satellite phone rang. The rocket was lost. As the SpaceXers began processing their shock, the satellite phone rang again. Kellie and the crew aboard *GO Quest* were given new coordinates at sea, and soon embarked on a very different kind of recovery mission.

| 8 |

# TRAGEDY AND TRIUMPH

*June 28, 2015*
Hawthorne, California

Seconds after the Dragon-bearing Falcon 9 rocket broke apart over the Atlantic Ocean, David Giger shouted into his headset, "Dragon is alive!"

In the decade since he joined the company straight out of graduate school, Giger had taken on management of the entire Dragon program, reporting directly to Musk. He watched the CRS-7 launch from mission control in Hawthorne not with a particular role, but rather providing a leadership presence. Giger could sense the Dragon mission team, mostly younger engineers, freeze up as video showed debris from the rocket showering back to Earth. A lot of the people involved in the hairy early flights of Dragon, including the C2 mission in 2012, had moved on to other positions at SpaceX, or departed.

"They were a great team, but I think everyone assumed it was over," Giger said. Unlike a lot of his colleagues, Giger had endured some trying times at SpaceX, including three failures of the Falcon 1 rocket. After the Falcon 9 shattered, Giger noticed that Dragon continued to send data back. It had separated from the rocket and was flying some thirty miles above the Atlantic Ocean.

The key to saving Dragon was opening its parachutes before it got too close to the ground. SpaceX had not anticipated such a contingency, nor planned to send commands to Dragon as it rode on the Falcon 9 rocket. But in an emergency, the Dragon control center could talk to Dragon using ground-based antennas. So controllers in California frantically worked to configure this communications system and command the two drogue parachutes to open. These are the small, precursor parachutes that stabilize the vehicle prior to deployment of Dragon's three main parachutes.

The command was sent, but nothing happened.

Dragon continued to dive. For about two minutes after the rocket's breakup, the plucky spacecraft faithfully relayed data. Then, less than a mile above the ocean, below the horizon from the Florida coast, the data stopped. The spacecraft and its 4,000 pounds of cargo plunged into the sea.

The problem is that opening the parachutes was not as simple as pressing a button. It required operators to send more than a dozen specific commands, in the correct order, to manually deploy the parachutes. In their haste—the operators had only seconds to act before the spacecraft fell too close to the ocean for its parachutes to have a meaningful effect—they had forgotten to turn on power for parachute deployment.

The unnecessary loss of Dragon was one lesson SpaceX took away from the failure of the nineteenth launch of the Falcon 9 rocket. In the aftermath of this CRS-7 failure, during long and intense meetings, Musk focused most of his time and energy on the rocket. But more than a few times, Musk would turn toward Giger and other Dragon officials and complain about its loss. "Dragon shouldn't be fucking stupid,"

Musk would admonish the team. "It should have saved itself." By the next Dragon mission, this emergency scenario would be baked into the rocket's launch procedures. If need be, Cargo Dragon could be saved.

The high-profile failure of the CRS-7 mission in June 2015 hit SpaceX hard. After five years of challenging work with Falcon 9 and Dragon, the company had started to hit its stride. The Falcon 9 won a string of lucrative commercial satellite contracts. NASA, too, invested billions of dollars in SpaceX to launch humans onboard Dragon one day. Then it all fell apart. Carrying cargo for its most important customer, SpaceX blew up its rocket in spectacular fashion, for all the world to see.

The loss of CRS-7 came at an inopportune time for NASA, as well. Half a year earlier the space agency's other commercial provider of cargo services, Orbital Sciences, lost a NASA mission when its Antares rocket exploded just above the launch pad. Critics of the agency's support for commercial spaceflight reemerged in Congress, underscoring the severity of the setback and raising questions about whether private companies could be trusted with human spaceflight.

## Acquiring hallowed ground

In the first three years SpaceX flew the Falcon 9 rocket, from mid-2010 to mid-2013, it launched just five times. This flight rate was completely unacceptable to Musk, and after introducing version 1.1 of the rocket he expected this cadence to increase substantially. He charged the Florida site director, Brian Mosdell, with delivering the capability of launching more than one Falcon 9 rocket a month.

But that was not Mosdell's only task. He also spent much of 2013 spearheading SpaceX's charge to obtain a lease for a second launch pad in Florida. This was to be none other than the most historic launch pad in the Western hemisphere, NASA's Launch Complex 39A.

Surrounded by swampland and rising just a few feet above the Atlantic Ocean, the sprawling 200-acre site includes the hallowed ground where Neil Armstrong and Buzz Aldrin took their last steps on terra firma before walking on the Moon. Later, dozens of shuttle missions also launched from there.

Following retirement of the shuttle in 2011, NASA no longer had any use for the launch site. The agency's inspector general characterized the pad as "unneeded infrastructure," and leasing the facility to a commercial launch company would offload millions of dollars a year in maintenance costs. SpaceX, already flying, seemed like the obvious choice as it sought a suitable site for Falcon 9 and Falcon Heavy launches and eventually crew flights for NASA. However, in the spring of 2013, another bidder emerged. Jeff Bezos held a reverent fondness for space history and sought to lease the site for Blue Origin and its New Glenn rocket.

To sweeten his proposal, Bezos offered to share the launch complex with SpaceX or another company. While this was a generous offer, Blue Origin did not have an orbital rocket in 2013, nor was it close to having one. Because of this, NASA awarded a twenty-year lease to SpaceX in September.

Musk was giddy after winning, having secured the landmark site. But he was also tweaked by Bezos's bid, which he believed was intended to block his access. In an email to *Space News*, Musk mocked his would-be competitor. "If they do somehow show up in the next five years with a vehicle qualified to NASA's human rating standards that can dock with the space station, which is what Pad 39A is meant to do, we will gladly accommodate their needs," Musk wrote. "Frankly, I think we are more likely to discover unicorns dancing in the flame duct."

Musk also mocked Blue Origin during internal meetings at SpaceX. He would say things like, "A company must really stink to call themselves

BO." History vindicated him. Blue Origin did not have an orbital rocket ready to fly from Launch Complex 39A within five years. Or ten. As of the writing of this book, the company's New Glenn orbital rocket has yet to make a single launch attempt. SpaceX, in the meantime, has launched more than 100 rockets from the old NASA pad.

Mosdell wrote a majority of SpaceX's proposal to win the lease for 39A, including all of its technical materials, its schedule, and its budget considerations. After the company acquired the lease, he fully understood all of the work entailed in demolishing the old shuttle infrastructure and building a launch tower capable of supporting the Falcon 9 and Falcon Heavy.

His relatively small staff included seventy-two full-time employees and about eighty welders and other contract hires working on a new transporter. They already were maxed out by Musk's push to launch monthly from SLC-40, putting in 80- to 100-hour weeks to keep up. Accordingly, Mosdell traveled to Hawthorne in January 2014 to meet with Musk and Shotwell about staffing up his operations for the build-out of Launch Complex 39A. During the meeting, Mosdell outlined what he believed to be a "lean" team to complete the design, procurement, building, and testing of the site. He then asked to hire sixty-eight people over the next six months.

Musk rejected it out of hand. The meeting spiraled downward from there, and Musk's parting words to Mosdell were, "Go home and work harder."

That hurt. As a leader, Mosdell keenly felt the demands he placed on his team with intense work schedules. Eventually they would burn out and leave, or the quality of their work would suffer. When he raised these concerns with Musk and Shotwell, they would always say the money was tight, and if Florida could just buckle down and get the next payment milestone, some of the burden could be lifted off the team. Mosdell's meeting in Hawthorne convinced him things really were never going to change. SpaceX would never enter a cruise phase. It would ever accelerate.

"There was not going to be a genuine effort to fix the problem," he said. "This was going to be a *shut up, go home, and color* kind of thing. And I decided I was no longer going to go to the Cape and tell my team to just keep going because things would change. Because I knew it was bullshit."

During his six years at SpaceX, Mosdell had delivered. He built the team that scrappily built the SLC-40 launch pad for about a tenth the cost of SpaceX's competitors. He worked as a launch director for six missions. They started with almost entirely manual launch operations, and by the end of 2013 reached the point where about 90 percent of the countdown was automated. Under Mosdell's watch, the launch site had a perfect record. He also led the campaign to win NASA's competition for its historic LC-39A site. He had laid the foundation for the success that ultimately awaited SpaceX in Florida.

But it had not been enough. He resigned.

Musk was not displeased. In his mind, Mosdell and the team at Cape Canaveral had not gone hardcore enough. Launching once a month was not enough. Musk believed Falcon 9 rockets should be flying from Florida once a week. In hindsight, this is a cadence the company would not reach for nearly a decade.

To fill Mosdell's role, Musk turned to one of his people. Ricky Lim had joined SpaceX in 2008, spending months on Kwajalein during the final three flights of the Falcon 1 rocket. He came of age during the early years at SpaceX, surviving the crucible of the Falcon 1 and near death of the company. He then helped out at Vandenberg, alongside Zach Dunn and Lee Rosen. In the wake of Mosdell's departure, Lim was asked to fill in as site director at Cape Canaveral for a few weeks. Three weeks would turn into six years.

The launch team at the Cape consisted of two halves. One was built from converts from the legacy rocket companies, including United Launch Alliance, where Mosdell had come from. The other half consisted of young engineers who had typically taken their first job with SpaceX. Musk felt Mosdell was too closely aligned with the legacies. Lim, he felt, would rally the younger blood.

And so he did. From mid-2014 onward, SpaceX started to hit a groove at the Cape. There were bumps along the way, of course, and eventually SpaceX would hire not sixty-eight additional employees as Mosdell had requested, but hundreds more for launch and pad rebuild work in Florida. Yet by April 2015, Lim and his team managed to launch a pair of missions—the sixth operational cargo mission for NASA and a Turkish communications satellite—within just thirteen days of one another from Florida. Eagerly, they moved to the next mission on the manifest, another cargo supply mission for NASA.

The seventh one. CRS-7.

## A painful, shocking moment

On the morning that CRS-7 lifted off, Lim set aside his site leadership role to serve as launch director. All went well for the first two minutes of the flight. However, at 2 minutes and 19 seconds, Lim started to hear chatter on the network about "data drops" from the second stage. He glanced up at a video showing the rocket far down range, a white streak climbing into an azure sky. While there was some sort of vapor cloud around the upper stage, the rocket's nine engines were burning just fine.

"It was surreal," Lim said. "What we saw in the data did not match what we were seeing. The first stage was still going. On the long-range cameras, it looked a lot like prior launches. I thought the second stage people were mistaken, to be honest."

He believed it might be a ground software display issue or some other minor problem. A second or two more clarified matters. The large white cloud expanded, engulfing the rocket. As it dissipated, the tracking camera showed a shower of debris as pieces of the rocket began tumbling back to Earth. A hushed silence fell over the flight control room.

Musk watched the launch from England, where he was celebrating his forty-fourth birthday. Seeing his rocket disintegrate was a lousy birthday present. His first call was to Dunn. After working for five years launching rockets, Dunn had taken over running SpaceX's propulsion department just a few months earlier and observed the launch from inside the Mission Control Center in Hawthorne. Dunn told Musk he did not have an immediate explanation for the failure and passed his phone to Jon Edwards, an engineer supervising the launch. Edwards said it looked as though some kind of pressure event had occurred on the second stage.

During past traumatic moments, Dunn had felt empowered to take immediate action. When a Falcon 1 rocket had started imploding during transport on a C-17 rocket over the Atlantic Ocean, Dunn had risked his life to climb inside the booster to open a pressure valve.

But this was different. "In this case I didn't really know what to do immediately," he said. "That was kind of the creepy thing about it. We knew we had to go and see if there was any debris in the ocean. But there was not a ton of immediate action other than starting to figure out what took place. It was a really painful, shocking moment."

Dunn walked out of mission control and back to the propulsion area of the office. He called his team together and gave a quick talk about what had just happened. Like Giger, Dunn was now one of SpaceX's veterans and had persevered through the Falcon 1 failures. He explained that launching rockets was a high-risk business, and that the way to get through this challenge was to work methodically to find out what happened, fix the problem, and fly again.

Musk returned to California and dove into the accident investigation with frenzied energy. At least daily, and often multiple times a day, he convened meetings in his executive conference room on the first floor of the company's headquarters. A black conference table dominated the middle of the room, around which nine chairs were situated in a large U. People presenting at the meeting filled the chairs. More engineers sat or stood around the sides of the room. Musk always stationed himself at one end of the table closest to the door, facing a wall that featured a large canvas print of the Falcon 1 rocket launching from Omelek Island, a cautionary tale of the company's desperate roots.

"The vast majority of the people at the company today have only ever seen success, and so you don't fear failure quite as much," Musk said after the accident, which he publicly described as a "huge blow" to SpaceX. "To some degree I think the company as a whole maybe became a little complacent."

Indeed, the company had grown ten-fold since its last failure in 2008. As the Falcon 9 had streaked through its first eighteen launches, some of the Kwajalein veterans like Dunn and Lim would discuss how few of the company's employees had experienced the pain of launch failure. By the time of the CRS-7 mission, only about 5 percent of employees had that experience, so they were somewhat numb to Musk's mantra that "only the paranoid survive." Before every Falcon 9 launch, Musk would email the entire company and say that if anyone had a serious concern about the upcoming launch, he wanted to hear it. And he was genuine about this. He wanted to hear risks and, if they sparked wider concern, act on them.

Because of SpaceX's growing size, downtime was expensive. With a staff of approximately 4,000 employees at the time of the failure, SpaceX needed to fly frequently to pay the bills. Personnel costs alone exceeded $70 million a month. So Musk pushed for a thorough, but rapid, investigation. During the CRS-7 failure analysis meetings he would listen as department leads shared their latest information on the investigation. Often he interrupted with a question, a comment, or a sharp critique. These were tense and frustrating and difficult discussions because the true cause of the accident remained a mystery for several weeks. Underlying it all was a sense of unease and uncertainty about the fate of SpaceX.

"Elon brought a tremendous amount of energy to those meetings," Dunn said. "You always have to bring your best with Elon, but in these sort of tense moments you had to be right there. One hundred percent focus. No bullshit. You better be able to explain what you know, what you don't know, take direction very crisply, and understand what he means with precision. If not, he pounced. You just had to be on your A-plus game. You had to be fucking great."

The responsibility for investigating the failure fell on one of Musk's oldest and most trusted lieutenants, Hans Koenigsmann. At the time Koenigsmann served as vice president of Flight Reliability and Mission Assurance, which essentially meant ensuring safe and successful launches.

"I feel like if I'm responsible for the risk, I'm also responsible for when it goes wrong," he said. "It took five months, and I worked around the clock. I took one weekend off, I think. Otherwise, I worked every single day of those months. I worked my ass off, and my team did, too."

Part of the challenge in understanding what had gone wrong is that the rocket had gone from flying normally to a conflagration in just 800 milliseconds. Koenigsmann's team knew almost immediately that the problem originated on the second stage, which had not yet separated from the first stage. But it had all happened so fast, in a fraction of a second. As a result, engineers and scientists from SpaceX and NASA had just 115 pieces of telemetry data (that is, measurements from various sensors onboard the rocket during that critical second when the upper stage failed). From this, they ultimately determined that the liquid oxygen tank had ruptured after a container containing pressurized helium broke free inside the tank.

Why are there bottles of helium in the LOX tank? As a rocket launches into space, it steadily burns propellant and oxidizer. As these fluids drain, helium gas is released to fill the vacated volume and maintain a downward pressure. This ensures that propellant and oxidizer continue to flow into the engines. The helium container is called a composite overwrapped pressure vessel, or COPV, because a strong fiber is wrapped around a metallic container to ensure its integrity. The more difficult question for Koenigsmann, therefore, became understanding why a COPV bottle containing helium had broken loose and struck the upper dome of the oxygen tank with catastrophic effect.

Eventually some structures engineers pinned the cause of the failure down to a small $4 part about the size of a Tootsie Pop. This stainless steel eye bolt, also known as a rod end, helped secure the COPV to the wall of the oxygen tank. It had broken during the rocket's ascent. Koenigsmann said these rod ends were rated to withstand 10,000 pounds of force, but one of them inside the ill-fated upper stage had broken under less than 2,000 pounds of force.

Nearly all rockets and spacecraft undergo a rigorous design process. The last checkpoint before a project moves into fabrication is known as "critical design review." During this review for the Falcon 9, SpaceX had required the use of a more expensive rod end, Koenigsmann said. This part costs about $50. Somewhere between this design review and the actual flight, however, a cheaper rod end had been substituted. This steel rod end cost significantly less and was manufactured by a casting process. This meant it was made by pouring steel into a mold, where it solidified, and then was ejected from the mold. The more expensive rod ends worked fine at cryogenic temperatures inside the liquid oxygen tank, but cast materials were more problematic when pulled under the force of tension as they could have unknown flaws hiding inside.

SpaceX tried to find the actual rod end that had broken, to ensure it had nailed down the original cause of the failure. Searching for a thumb-sized part fifty miles off the coast of Florida, hundreds of feet below the surface of the ocean, proved to be a rather Don Quixote–like quest. The company even hired a remotely operated submarine to look for wreckage. In the process, it found long-lost hardware from the Apollo and Space Shuttle Programs, but no Falcon 9 parts. However, Koenigsmann was confident that a cast rod end was the culprit because SpaceX tested similar parts from the same purchase order and found they were subject to breaking under cryogenic conditions.

So why had SpaceX switched to a cheaper cast rod end? Musk had instilled a culture of always looking to cut costs. *Someone* decided that a $50 rod end was too expensive and would be substituted with a cheaper part. Every strut on the rocket used rod ends, so there were hundreds of them on the vehicle, meaning this single change saved more than a thousand dollars on a launch vehicle. In their zeal to control costs, Musk and his lieutenants made such decisions thousands of times. Had Musk not been so judicious about costs, the price of a Falcon 9 would have ballooned. And in nearly every case the approach worked—except this one.

In the fall of 2015, Koenigsmann wrote a detailed report about SpaceX's findings and presented it to NASA and the Federal Aviation Administration. The report concluded that "material defects" were the most probable cause for the broken rod end. This essentially put the blame on the rod end's supplier. NASA's own, independent findings leveled a harsher judgment more directly on SpaceX. NASA attributed the failure to a "design error" by SpaceX. The space agency also said SpaceX's quality control process should have identified the substandard rod ends before they were installed on the rocket.

"The implementation was done without adequate screening or testing of the industrial grade part, without regard to the manufacturer's recommendations for a 4:1 factor of safety when using their industrial grade part in an application, and without proper modeling or adequate load testing of the part under predicted flight conditions," the report stated.

In other words, NASA said SpaceX had messed up, not the supplier. Koenigsmann said he accepts that SpaceX should have done a better job screening the rod ends. But he said the supplier deserves blame as well. "SpaceX *and* the supplier screwed up," he said.

Koenigsmann still keeps one of these rod ends in a desk drawer at home, to not forget the lessons of CRS-7.

Despite disagreeing over the fundamental cause of the supply mission failure, NASA and SpaceX continued to work together well. When Dragon sank into the ocean, NASA lost $118 million in cargo, including a crucial docking adapter needed to enable future astronaut missions to the space station. The failure also increased the U.S. space agency's reliance on Russia. For several months, the only means America had of getting its astronauts to the space station, and feeding them, came via a pair of small spacecraft designed during the Soviet era that launched from Kazakhstan.

Publicly, however, NASA did not chastise SpaceX for these problems. Rather, its officials remained supportive. During a U.S. Senate hearing in 2016, when some elected officials would have celebrated an opportunity to lambaste the company, NASA's chief of human spaceflight stood up for SpaceX when asked about the failure.

"They turned around very quickly," the official, Bill Gerstenmaier, told Congress. "Within a matter of days, they were actually in a test facility on the ground testing the failure that they thought had occurred. That getting into test was much faster than I could have ever done on a NASA side. By the time I would have had the ability to get contracts written and done the proposals and put the test sequence in place, it would have been a half a year."

SpaceX, in turn, sought to make NASA whole for its losses. A few months after the accident, the company quietly agreed to fly five future cargo missions, its sixteenth through twentieth flights to the space station, at discounted prices. SpaceX also increased the amount of cargo each Dragon mission would carry, giving NASA more bang for its buck.

This assumed, however, that SpaceX could get the Falcon 9 rocket flying safely once again.

## Musk's risky decision on liquid oxygen

Even had the cargo mission not failed, the Falcon 9 still faced months of downtime during the second half of 2015. Musk had decided the rocket needed another significant upgrade, to version 1.2, which later became known as Falcon 9 Full Thrust. This represented a massive evolution to the rocket's capabilities. Drone ship landings were essential to making the economics of first stage reuse work, but they were not the only step. SpaceX also needed to squeeze every ounce of performance out of the rocket. No part of the Falcon 9 was spared a ruthless revision, and in the end SpaceX engineers produced a new machine that increased the lift capability of the Falcon 9 by nearly one-third.

The propulsion department designed an upgraded version of the Merlin 1D engine that raised the thrust of each engine by about 15 percent. The structures department built a lighter rocket that was easier to manufacture. And all of the lessons learned from the Grasshopper program and landing attempts in the Atlantic Ocean were poured into the design of the new rocket legs and control systems.

However, the real linchpin of the upgrade involved a technology known as propellant densification, or squeezing as much fuel as possible onto the rocket. This sounds wonky and wholly uninteresting—but it is not. The science and engineering of super-chilling rocket fuel is fascinating, and its implementation tremendously risky. Within a year of seriously starting work on densification, SpaceX would blow up a rocket, destroy a launch pad, and lose a $195 million Israeli satellite. Some former Apollo astronauts viewed SpaceX's approach as so dangerous they urged NASA to never let its astronauts fly on rockets fueled this way. Yet Musk did not flinch. He knew the risks, he accepted the risks, and ultimately he and SpaceX conquered the risks.

Densification, however, was a tremendous challenge heaped on SpaceX at the same time the company's employees were scrambling to recover from the loss of the CRS-7 mission, satisfy NASA's concerns, and work through the finer points of landing first stages.

"Elon grasped the essentials of the reuse problem," John Muratore said. "He kept telling us we've got to get more performance. We've got to get the liquid oxygen colder. He just kept driving us." Somewhat understatedly, Muratore added, "It was quite an intense time."

SpaceX densified both oxygen and kerosene, but since the former had to be chilled to much colder temperatures, it was far more difficult to handle. Oxygen is the most abundant element on Earth, and essential for life. Humans cannot breathe without it, and in our bodies it chemically reacts with molecules from food to produce energy. Similarly, this process of oxidation occurs when oxygen combines with a fuel. Firewood, for example, cannot burn without oxygen. And so oxygen is an essential component of producing combustion within a rocket engine. In fact, most rockets burn more oxygen than fuel on the way to orbit. Onboard the Falcon 9, in terms of mass, there is more liquid oxygen than kerosene fuel.

Musk reasoned that by packing more liquid oxygen into the rocket, it could get better gas mileage. He was certainly not the first engineer to think about forcing oxygen into a denser state and thereby increasing the amount that a rocket's tanks could hold. NASA had previously studied propellant densification over the decades.

Recently the agency had dismissed it, yet again, for the Constellation Program. But this decision was not based solely on physics. Rather, it was due to politics and rivalries between the agency's field centers. Marshall Space Flight Center, in Alabama, already had its bread and butter with existing propulsion technology. And NASA management had little appetite for the exploding test articles that would necessarily accompany densification development. Neither of these were barriers at SpaceX, which could afford to fail—and indeed publicly celebrated its test failures as evidence of pushing beyond the bleeding edge.

By densifying liquid oxygen and kerosene onboard Falcon 9, SpaceX could squeeze an extra 8 to 10 percent of performance out of the vehicle. This was not trivial. It meant carrying two more tons of payload to orbit. This was extremely important for a reusable rocket, which was paying a significant mass penalty for returning to Earth due to its landing gear and other added components. For economic viability, therefore, Musk believed densification was as important as drone ship landings. If he could accomplish both, the Falcon 9 could truly be the world's first twenty-first-century rocket—reusable, high-performing, and cost-effective.

So how does one densify oxygen? One way is to use liquid oxygen instead of gas. Liquid oxygen has a pale blue, ghostly color. It condenses at –297.33 degrees Fahrenheit (–182.96 degrees Celsius), far, far colder than the coldest temperature ever recorded on Earth in Antarctica. It is colder than even the darkest areas of the Moon that never see sunlight. This makes working with liquid oxygen difficult. However, the upside for rockets is worth it: liquid oxygen is 1,000 times more dense than gaseous oxygen, so most rockets use liquids.

What Musk wanted to do was make this liquid oxygen still more dense by chilling it down, almost to a solid. This is basic chemistry, as the cooler a substance gets the more its constituent molecules slow down, thereby bringing them slightly closer together. So the colder SpaceX could get its liquid oxygen, the more that could be packed onto the rocket.

This is how one day, in 2015, Muratore and another engineer named Vincent Werner found themselves on the phone with the National Institute of Standards and Technology, a Maryland-based agency that is the world leader in measuring physical properties. Werner and a handful of SpaceX engineers had been poring over tables published by the agency that showed the various temperatures and pressures at which oxygen, nitrogen, and liquid air—a mixture of mostly oxygen and nitrogen—turned into solids.

"We called them because they had generated the tables," Muratore said. "And they were like, 'You know, guys, these were extrapolated tables. Nobody's ever worked down here before. The tables are approximately accurate, but they could be off by a degree or two, or a psi or two.'"

SpaceX was not just looking to experiment with liquid oxygen at its coldest temperatures; it planned to produce vast quantities of the stuff. For a single rocket launch, the company needed to make hundreds of thousands of gallons. The actual work of producing densified oxygen fell to a small team of about eight engineers in Cape Canaveral, including Phillip Rench.

He seemed an unlikely hire for SpaceX. Rench earned a degree in mathematics from Southern New Hampshire University, which is not known for aerospace greatness. Rench then spent nearly a decade working at SeaWorld in Orlando where he performed an odd assortment of jobs, from underwater maintenance to fixing amusement park rides. While working at SeaWorld, Rench discovered a knack for devising solutions to challenging problems. In 2010, a veteran trainer at the park, Dawn Brancheau, was dragged to her death by a killer whale named Tilikum while gently rubbing the creature. After the incident, Rench helped build a giant submersible floor that lifted the whales out of the water, to make it safer for trainers to interact with the creatures. It was Rench's first time working with complex valves and other components used in control systems.

After seeing a promotional video that depicted a Falcon 9 rocket launching and landing in Florida, Rench was blown away. So he applied to SpaceX and was hired early in 2014 to help modify Launch Complex 39A. Rench watched the fateful CRS-7 launch from the vantage point of this pad, alongside the other engineers, technicians, and interns working on the old NASA site.

"Everyone was super depressed," he said. "But the next day we came back at 150 percent, with energy and passion. You know the five stages of grief? Yeah, we went through that really quick."

Engineers at McGregor had performed preliminary densification tests, and some of the early work in Florida was led by Brian Childers and Gavin Petit. Rench worked with a team that included Petit, David Ball, Chris Wallden, and others. Because the Florida crew had no practical experience with super-chilled oxygen, they more or less just started connecting equipment and seeing what happened. SpaceX used liquid nitrogen to chill liquid oxygen because the colorless gas turns into a liquid at –320 degrees Fahrenheit (–196 degrees Celsius), below that of LOX. To further chill the liquid oxygen, the engineering team flowed it through a pipe, around which was wrapped a coil of tubing filled with liquid nitrogen. The two never mixed, but the warmer LOX would shed heat into the liquid nitrogen. As this heat moved in, some of the warmer bits of nitrogen started to boil off. SpaceX used very powerful vacuum pumps to suck this heat away. Over time, as the pressure dropped, the temperature of nitrogen fell below –340 degrees, and the liquid oxygen followed. They could not go much colder, as nitrogen freezes at –346 degrees Fahrenheit (–210 degrees Celsius).

Rench loved the work. He had spent years working on valves and other systems to control the flow, temperature, and pressure of liquids. When they were pushing liquid oxygen to its extreme, it was not so different from SeaWorld. Over the course of a few weeks, he and the other engineers developed procedures by which this super-chilled LOX could be made and stored in a large, insulated tank at the LC-39A launch pad. They worked in pairs, for eight-hour shifts, seven days a week. The nights were eerie, with a soundtrack from purgatory.

"Liquid oxygen does not want to be densified," Rench said. "Densification makes this low, horrible growl. When we first started densifying LOX, the Praxair delivery drivers would be pumping the warm LOX into the sphere and it would make all kinds of crazy noises. They were getting nervous to be around it, and these are people who have worked with liquid oxygen for pretty much their entire lives."

NASA had been skeptical about SpaceX's plans for densification at Launch Complex 39A, so it asked for a demonstration. After Rench's team delivered and NASA signed off, the fluids team at the launch pad started stripping the parts and pumps away from the LOX chiller system. They were needed at SLC-40 to make the densified propellant for the debut flight of the Falcon 9 Full Thrust.

## SpaceX desperately tries to save Christmas

After two failed drone ship landings, Musk felt ready to try landing on land. This had a major advantage over the ocean, as the rocket need not contend with high seas. Ground was ground—flat and unmoving. But there was a major disadvantage, too. In returning the Falcon 9 rocket to land, it would fly near cruise ships in Port Canaveral, the National Reconnaissance Office's multibillion-dollar Eastern Processing Facility, and numerous other launch pads and valuable assets.

SpaceX acquired an old Cape pad, Launch Complex 13, in February 2015 for the purpose of returning the rocket. Trip Harriss, who had been with the company since its days on Kwajalein and the Falcon 1, now bore responsibility for Falcon recovery efforts and led the build-out of Landing Zone 1. He and Bala Ramamurthy also worked to convince the Range commander that SpaceX should be allowed to aim rockets at the Air Force station, a first.

"As the Range commander you're used to rockets going away from you," said General Wayne Monteith, who commanded the 45th Space Wing at Cape Canaveral from 2015 to 2018. "So when you see one that's 180 feet tall and coming back, as the person responsible for the safety of everyone on that installation, you start to get a little worried. Your career dissipation light starts blinking."

Harriss and SpaceX provided data to convince Monteith and other Air Force officials of the project's safety. It helped that the ocean landings, although they were not successful, had come close to hitting the drone ship. So Monteith felt confident that if SpaceX damaged any property at the Cape, it would be the company's own equipment. SpaceX also demonstrated that the vast majority of the booster's return flight profile was over water, only coming ashore at the landing site in the final seconds. That way, if something went wrong, a destruct signal could be sent to the first stage before it threatened anything on shore.

Before he signed off on a landing attempt, however, Monteith had to convince his supervisors the plan was safe. As the weeks ticked down toward SpaceX's return to flight, opposition started to get louder from the National Reconnaissance Office, which was concerned that vibrations from the rocket's sonic boom—as it slowed from supersonic to subsonic speeds—would damage the delicate work being done in its payload processing facility. Monteith also weighed similar concerns from his own team.

Range safety analysts predicted the Falcon 9 flyback would produce a sonic boom comparable to the major 2013 Chelyabinsk meteor event in Russia, damaging buildings and homes in the Cape Canaveral area and causing widespread damage. There was little data to refute these claims, which came as part of a lengthy and official-looking 100-page report defending the analysis. Alongside those claims came a stark warning that the United States would lose assured access to space, possibly for years, due to damage of critical launch facilities.

Why was there such caution? Unless the military is in the midst of a war, it is a risk-averse operation. Asses are on the line if there is a screwup. Monteith knew the buck stopped with him and that by making the call to allow SpaceX to land at Cape Canaveral, it was his particular ass in the line of fire.

"During a meeting, a commander's call, I stood up and said I believed this was the right thing to do," Monteith said. "In doing so I understood that if anything went wrong, I would be fired."

In early December, SpaceX received a green light from the Air Force to not just launch a missile, but to bring one back to the station. This is pretty remarkable, as SpaceX was flying a brand-new version of its Falcon 9 rocket, which was returning to flight after a launch failure, with densified propellant onboard for the first time. This was a brave call by the brigadier general.

Predictably, the run-up to the launch was chaotic. After SpaceX solved the rod end issue with NASA and the Federal Aviation Administration and obtained permission from the Air Force to dive-bomb its rocket back to the Cape, it still had to refine new procedures for densified oxygen.

One challenge with the densified oxygen was the inability to "recycle" a launch attempt if there was some technical or weather problem at the appointed time for liftoff. Once the super-chilled propellant was loaded onto the rocket, SpaceX had minutes to launch, or the liquid oxygen would become too warm. Although there was spare oxidizer in the LOX ball, offloading the warmed liquid oxygen from the rocket to this storage vessel would spoil the colder oxygen there. Dumping all the rocket's LOX was not an option, as this would damage pipes and other launch site infrastructure.

As launch director, Lim also kept a concerned eye on the calendar. SpaceX had targeted the night of December 21, 2015, for the return to flight launch. The rocket would loft eleven satellites for the telecommunications company ORBCOMM into low-Earth orbit, with a total mass of about 4,500 pounds. This was a light enough load for the Falcon 9 to have plenty of spare fuel to return to Landing Zone 1. Everyone had worked intensely to get ready for the launch and was counting on a few days off over the holidays. Many talked of quitting if they did not get a break soon.

"We were desperately trying to save Christmas," Lim said. "Our employees had been working months on end, and I worried that about a third of them might leave. If we scrubbed, and plowed through the holidays, it would just have been murder."

Once again, Lim directed the launch from inside the company's control center about eight miles from the pad. In the early years there were two principal leaders during a launch, the director and the chief engineer. This created a tension on launch day as the director served as the "gas pedal" and the chief engineer a more cautionary role as the "brake." Koenigsmann typically served as chief engineer of launch, but due to his focus on the CRS-7 failure, he delegated the role to Robb Kulin.

Koenigsmann and Musk watched proceedings from inside the control room. Both men felt the gravity of the moment. The countdown was tense. Then, in the final minutes before liftoff, scheduled for 8:29 PM local time, a camera inside the interstage area between the first and second stages showed drops of a pale blue liquid dripping down. This was a novel problem due to working with densified propellant for the first time, and it might indicate a number of bad things. Kerosene leaking might result in a fire. Liquid oxygen could lead to an explosion. Reviewing data and video, the launch team determined it was probably "liquid air," or air that had been cooled down to cryogenic temperatures by the frigid tanks. Hastily, the launch team discussed whether to scrub and investigate the leak.

At T-1 minute, Koenigsmann turned to Musk. "You've got to make a decision," he said.

Musk looked back at Koenigsmann. Almost invariably, Musk made his decisions with strength and self-confidence. He commanded. Others obeyed. But in this instance, with everything on the line for SpaceX, he responded casually, almost dreamily. "Well, I guess we're going," Musk said.

And so they were.

The upgraded version of the first stage performed perfectly. After dropping off the second stage, the rocket burned for home, dropping out of the black night down to the Florida coast. Nearer to land, from the vantage point of the launch control center, the rocket disappeared behind the tree line with a spectacular orange glow and a huge cloud of dust.

Then, there was a huge, building-shaking blast.

"That scared the shit out of us," Koenigsmann said. He and Musk thought the rocket exploded. In response to the deafening boom, Musk's countenance sank, despondent and disappointed.

Someone on the launch team suggested they check the video feed from the landing site. It told a happier story. The Falcon 9 rocket? It was there, standing upright on the landing pad, smoking in the mild Florida evening.

They had been fooled by the reentering rocket's sonic boom, which had been delayed a few seconds traveling to the launch control center. The room erupted in applause and cheers.

Musk's mood reversed entirely, becoming delirious with joy, absolutely smashed full of happiness and pride for persevering long enough to see this moment. His faith in bringing rockets back from orbit and landing them, so often questioned for so long, had been validated. Like a kid in the candy store, he kept pressing Lim and the launch team to go out to the landing pad and see his beautiful rocket. Three different people who had been with Musk for years said they had never seen him happier than he was that night.

SpaceX had negotiated range safety protocols with the Air Force in the event of a landing. The rocket still had explosives onboard, including TEA-TEB ignition fluid, the rope-like flight termination system, as well as liquid oxygen and kerosene. A safety team had to secure the rocket first. But in less than an hour, Musk, Koenigsmann, and others, including Kulin, Harriss, Shana Diez, and Lee Rosen, donned hard hats to go running and skipping and dancing across the landing pad. As they danced about, they noticed there had been no apocalyptic meteor event, nor property destruction of any kind. Even the basic windows in an office trailer by the landing site bore nary a scratch. The launch of the new rocket and its unprecedented landing were a complete success.

"It's hard to describe how epic this comeback was after our first Falcon 9 launch failure," Koenigsmann said.

As he, Musk, and the others marveled up at that sooty rocket, illuminated by flood lights beneath dark and starry skies, they must have wondered if this moment, on this night, in this world, could ever be eclipsed.

## "It just felt so massive."

They were pretty excited back in Hawthorne, too. As the rocket touched down, hordes of employees crammed into the factory floor just outside mission control started chanting, "U-S-A! U-S-A! U-S-A!" A raucous celebration ensued.

And why not?

The four thousand employees of SpaceX had wrought nothing short of a miracle in the six months preceding that night. The company worked on four separate, massive projects in parallel, packing their final exams into that single launch. Riding onboard the Falcon 9 rocket in late December were the company's return to flight mission, a significant upgrade to the Full Thrust version, an unprecedented oxygen densification program, and the first landing. They saved Christmas, to boot.

The historic ORBCOMM launch and landing delivered one of the most cathartic and breathtaking moments in SpaceX history. I do not believe it is possible to overstate the significance. With its fate on the line, the company roared back from a terrible and financially disastrous failure. And, on the very same flight, SpaceX accomplished something no company, or country, had ever done before. Until then SpaceX had followed in the footsteps of NASA and others in launching rockets, flying satellites into space, and landing spacecraft in the water. Sure, it did so in cheaper and innovative ways. But these were well-trodden paths. No one had ever, ever launched an orbital rocket and landed it back on Earth minutes later.

Until that night.

Catriona Chambers came to SpaceX in early 2005 as an electronics engineer. Within months on the job, she picked up responsibility for the Merlin engine computer on the Falcon 1 rocket. On that small rocket's very first launch, there was a sensor that measured atmospheric pressure. After reaching space, the first stage would descend back to Earth, and when the sensor detected a thickening atmosphere, it would command deployment of a parachute. She and everyone who worked on the rocket knew this was preposterous. The rocket would probably never survive, and the parachute would be practically useless. But Musk pushed hard for reuse from the very beginning of SpaceX. Now here she was, almost eleven years later, observing it actually happen. As director of avionics, she watched with her team as the first stage landed, feeling the weight of history as she hugged and high-fived her friends.

"That was the point where it really sunk in that we had been working on this for so long," Chambers said. "It just felt so massive, and I was so excited. And then I realized I needed to calm down." She was eight months pregnant, after all.

Like a lot of SpaceX employees, Zach Dunn felt both exhilaration and relief at the launch of the ORBCOMM mission. He had taken over the propulsion department in February, with the aim of completing the Merlin engine upgrades for the Full Thrust version of the Falcon 9. Within the first couple of weeks on the job, two engines blew up. Then the CRS-7 launch failed, and Dunn was thrown into the tortuous investigation. Finally came the arduous campaign to ready the new rocket and update the launch site for densified propellant.

This gave the propulsion team fits right up until the launch date. On December 18, the company had to abort three separate attempts at completing a static fire test. The launch team was still learning, on the fly, how to load and offload super-cold liquid oxygen, when Musk came into the control room. Invariably, his presence raised the level of tension and urgency. Dunn explained to Musk that by the time the rocket was ready to ignite its engines, the propellant was warmer than the engines were expecting.

Musk told him to run the test anyway.

"My engine team was telling me this was not the right thing to do," Dunn said. "That we weren't going to get the data we needed from the test. The pressure from Elon was just absolutely intense."

On launch day, Dunn sat next to Shotwell in Hawthorne's mission control. As soon as the booster touched down, Shotwell leapt to her feet, joining the merriment. After a few minutes of also celebrating, Dunn left and walked across the factory floor to the propulsion area. About five dozen engineers were there, almost all of his propulsion department.

"It had been a hard fucking year," Dunn said. "This was my hardest year at SpaceX, leading propulsion, going through those failures and trying to keep the team together, and the pressure of getting back to launch. It pushed my leadership and technical abilities to their limits. My interface with Elon was more direct and more intense than it had been before. It had taken a toll."

As Dunn walked toward his desk, the other engineers, one by one and then in a rush, stood up and applauded him. A standing ovation. It was completely unexpected. Dunn had come into the department as an outsider, having led pad operations at Vandenberg. There were a lot of egos and a lot of brains in SpaceX's propulsion department. During the preceding ten months Dunn had fought with this team as well as for this team. He'd won some. He'd lost some. But after that night he was no longer just their leader. He was one of them.

"Man, I've never felt better in my life," Dunn said. "It felt incredible to experience that after the hardest fight that I've ever had professionally."

| 9 |

# F-SQUARED AND THE AMOS-6 DISASTER

*February 2009*
McLean, Virginia

Matt Desch thought he might be able to squeeze a few extra bucks out of Elon Musk.

So after summoning his best negotiating skills, Desch telephoned the SpaceX founder in early 2009. The chief executive of a satellite company named Iridium, Desch sought to expand his space-based communication network, and as part of that he needed to close a major deal with SpaceX for eight launches of the Falcon 9 rocket.

He knew well the importance of this contract to SpaceX, as it would be the launch company's largest commercial agreement to date. Moreover, Iridium's confidence in the rocket startup would send a clear signal to the rest of the commercial satellite industry that this company, which had put all of one customer into orbit at the time, was the real deal.

Iridium's negotiating team had talked the price for seven launches down to $492 million, but they felt there might be a little more wiggle room. For a final push they turned to Desch, the boss. He should call Musk, they urged, and extract an additional $10 or $15 million. As he dialed Musk from his office in Virginia, Desch felt as though he had some leverage because of the deal's significance for SpaceX.

"The contract is a little expensive," Desch said, after greeting Musk. "I think it could be a little less, Elon. I'm feeling like $480 million would make more sense."

Musk paused for a moment, perhaps thinking about his response. While the contract was a landmark deal for SpaceX, Musk also understood the value of the Falcon 9 rocket to potential customers. He did not know the exact offers his launch competitors made to Iridium, but Musk was confident SpaceX had bid the lowest price to put more than five dozen Iridium satellites into low-Earth orbit. Desch had to be bluffing.

"Matt, honestly, no offense," Musk replied. "We love your team. We love working with you guys. But I have other customers who would be happy to launch at this price."

To launch the Iridium NEXT constellation, Desch's negotiating team had gone to launch providers around the world, in Russia, Europe, and the United States. Some of these were state-run enterprises and some private businesses. But except for SpaceX, all were traditional companies, accustomed to charging high prices for the rare commodity of launch service. SpaceX had bid under $500 million. Iridium's next lowest offer was $1.2 billion. Desch could not risk messing this up. So he simply replied that he was ready to close the deal. As is.

"Great," Musk replied. "I look forward to it."

Years earlier, Iridium had been a poster child for monumental failure in space. At the end of the twentieth century, with financial backing from Motorola amounting to billions of dollars, Iridium launched a satellite service that enabled telephone calls via satellite phones. But after a number of management missteps, the company filed for bankruptcy. The court erased $4 billion in debt, and the U.S. government agreed to become a customer to help save Iridium from having to deorbit its satellites. A private group of investors took over the company.

Several new CEOs had come and gone by the time Desch, a telecommunications executive, was hired in 2006. Iridium was cash flow positive by then, but its original satellites were aging, with perhaps five to ten years of useful lifetime left. Desch pitched his board of directors on a replacement network of dozens of satellites, and a year later took Iridium public to raise money. The most critical part of the $3 billion plan was manufacturing of the satellites, and Iridium chose Europe-based Thales Alenia Space over Lockheed Martin, because Thales offered to help raise additional financing from banks backed by the French government. The meeting to woo these bankers was set for mid-June 2010, in a conference room at the swanky Four Seasons Hotel George V in Paris. This establishment features oversized suites that overlook the Eiffel Tower, and the price for such a room with two twin beds was about $6,000 a night in 2023.

As its schedule slipped, the first Falcon 9 launch ended up taking place just a week before Desch's meeting with European investors. The timing was fortuitous, because the investors had serious questions about the launch side of the deal with such an unproven company. Was SpaceX's price too good to be true? What if the Falcon 9 failed, like other commercial rockets before it? Desch prepared ahead of the meeting for a failure on the Falcon 9's debut, planning to explain that investors should not worry about a problem on a new rocket's first flight. But then the Falcon 9 shot into orbit, right on target. Gwynne Shotwell swooped in with video of the launch, which she exultantly shared with Desch's European investors.

"Thank God the first launch of the Falcon 9 was successful," Desch said. "I had forty finance and insurance people in a meeting room, and they were just mesmerized by the first videos of the launch of a rocket that nobody had really seen before. The timing was perfect."

Over the next six years Iridium worked closely with SpaceX to understand changes to the Falcon 9 and ensure its satellites would fit snugly inside the rocket's payload fairing. Increasingly, Desch and his launch team grew comfortable with SpaceX as it matured from a scrappy startup to a reliable launch company. By late August 2016, Desch believed his nearly decade-long dream of launching a next-generation satellite constellation was about to come true. He had bet his career, and the future of a multibillion-dollar company, on this plan.

Now there was just one more launch, an Israeli communications satellite, in line ahead of Iridium.

## "I'm the only person who calls a scrub."

Despite the stunningly successful ORBCOMM mission in late December 2015, the Florida launch team continued to struggle loading liquid oxygen quickly enough in the early months of 2016. It was during this period that long-time NASA veteran John Muratore emerged to help manage these challenging operations.

Balding, bespectacled, and gregarious, Muratore had a storied, nearly thirty-year career at the space agency. His assignments included managing the X-38 experimental space plane program, working as a flight director, and leading the space shuttle's engineering program. He left in the aftermath of the *Columbia* accident, when he disagreed about the orbiter being safe to fly again. After teaching for a few years, Muratore joined SpaceX in 2011 to work on government contracts, including certification for SpaceX to fly Dragon to the space station.

Muratore was not a typical NASA employee, and this helps explain how he fit into SpaceX so well. As demonstrated by his opposition to the shuttle's return to flight, Muratore was outspoken and unconventional in his thinking, and amenable to SpaceX's culture of thinking outside the box. Given his ample experience with the space shuttle, in 2015 Muratore was assigned the task of rebuilding its old pad, Launch Complex 39A. Having overseen shuttle launches from there, Muratore would supervise construction of a brand-new facility where SpaceX would fly astronauts aboard the next generation of American spaceflight vehicles.

After arriving in Florida, Muratore started to fill in for Lim as launch director on some Falcon 9 flights. The next mission after ORBCOMM was SES-9, a satellite launch for the Luxembourg telecom company. The rocket was ready to go late in February but scrubbed due to the liquid oxygen not being cold enough. The company tried again the next night, calling a halt with less than two minutes before liftoff due to difficulties loading super-chilled propellant.

After a three-day hiatus, SpaceX made a third attempt on February 29, 2016. This time, less than a minute before liftoff, Range safety officials called a halt when a boat strayed into restricted waters. A military helicopter chased the vessel out of the area, and the launch team scrubbed for the day. Given the demands of loading liquid oxygen in a timely manner, it was not realistic to recycle the rocket for another launch within the three-hour window. Lim and the chief engineer for the mission, Robb Kulin, called Musk to tell him what happened. His response was furious. "I'm the only person who calls a scrub," he screamed. He told the launch team to try again within the allotted window.

SpaceX had never attempted this before with densified propellant and had no procedures in place. The control room turned chaotic as the engineers started making up procedures on the fly, offloading warming liquid oxygen, planning to replace it, and asking Range officials for a new launch time. Amid the disarray, the liquid oxygen in the second stage started to expand, dangerously increasing pressure on the tank walls. "In retrospect, we nearly blew up the entire launch pad," Kulin said. The team had missed a critical valve command. A trickle of helium bubbles, used to mix the densified liquid oxygen and prevent water hammer, missed being turned off. These bubbles were sucked into the engine. The Merlin had to be replaced. Finally, five days later, the rocket launched.

All the delays left Musk in a very unhappy mood. After the SES-9 mission finally took flight, he called the entire launch team into a meeting, with teams from Cape Canaveral and Vandenberg joining by video. Lim instinctively knew it would not be a good meeting. He and several others bet on how many "F-bombs" Musk would drop, writing their guesses on a dollar bill with a magic marker. Lim guessed eight and won.

During the meeting, Musk reiterated that the launch site should never be the cause of a delay. Airports, he said, operate twenty-four hours a day without teams of engineers running around to solve problems fueling jets. As he spoke, Musk grew increasingly intense. And then he laid down the law.

"From now on, all I want out of you guys is fucking flawless," Musk said. Then he repeated himself.

At the Cape, Muratore instructed his launch pad teams at both Florida launch sites to take Musk at face value. "I told everyone that we were going to embrace F-squared," Muratore said. And so they did. One month later, without any delays, a Falcon 9 rocket launched its first Dragon mission after the failure nine months earlier. This launch of the CRS-8 mission in early April 2016 carried nearly 7,000 pounds of supplies and hardware to the space station.

Nine minutes after liftoff, however, no one really cared about the launch anymore. All eyes were on the *Of Course I Still Love You* drone ship a few hundred miles off the coast of Florida. SpaceX had attempted to land on a drone ship four times, and although they had come close to the mark, the rocket had always slammed into the boat or landed and subsequently tipped over.

Musk followed the landing attempt alongside Dunn in the front room of the launch control center at the Cape. Two years earlier, over Easter weekend, they had frantically chased after the possibility of finding scraps from the wreckage of a rocket in the ocean. Now they had a chance to land their prize at sea, intact. Musk, Dunn, Koenigsmann, and the other engineers at the Cape watched video relayed from an aircraft, leaning forward in their seats. The last seconds of the sooty rocket's descent to the deck of the drone ship were very slow, agonizingly so.

Upon reaching the deck, it bounced slightly.

"It touches down and you're thinking it's going to fall over," Dunn said. "But the smoke clears and there it is. It was total elation. Elon jumped up. We high-fived and he gave me a big hug. It was a magical moment. It blew my mind."

Hours later, Musk was still giddy. "We're a little bit like the dog that caught the bus," he said at a post-launch news conference. SpaceX was now in possession of a rocket it had not entirely expected to catch.

I remember watching the landing live from home, mesmerized. For years I had felt sorry to have missed seeing the Apollo Moon landings; I was born just months after the final one. But watching this rocket come back to Earth and land at sea was science fiction becoming fact before my eyes. I no longer felt sad about missing the feats of Apollo. I grew truly excited about being alive now, to see where this would lead us in spaceflight.

Musk had thoughts as well. "I think it's another step toward the stars," he said. "In order for us to really open up access to space, we've got to achieve full, rapid reusability. It will take us a few years to make that smooth and efficient. One day we'll hose the rocket down, add the propellant, and fly again."

That day may yet come. But in the near term, Musk's primary goal was increasing the launch rate. There were so many customers, like Matt Desch and his Iridium satellites, waiting to fly. And as spring turned to summer that started to happen, with the Cape popping off missions left and right. Lim, Muratore, and the rest of the Florida launch team conducted five launches in four months, marking the first time SpaceX achieved a sustained cadence above a mission per month. They, and the rest of the company, were delivering for Musk.

"Everyone started to believe in F-squared," Muratore said. "And then the accident happened."

## Anomaly on the pad

Once loaded with chilled propellant, a Falcon 9 races against time. Every second it delays, the warmer the propellant onboard becomes, and the lower its overall performance. Understandably, then, Musk pushed his operations teams to go faster, and faster, and then faster still.

Why? Think of it like this. Your rocket is happily basking in the Florida sunshine, near the beach. It's a lovely summer day. The temperature is about 90 degrees. It sounds delightful—unless you're a batch of freshly brewed super-chilled liquid oxygen. In that case, it is freaking *hot* outside. The temperature differential between the inside of the rocket and the outside is something like 430 degrees Fahrenheit. Think of how fast ice would melt if you set it in an oven that hot. In fact, one evening my daughters and I decided to try this experiment. We set our oven's temperature to 462 degrees Fahrenheit (240 degrees Celsius, or gas mark 9). We placed an ice cube, frozen by definition at 32 degrees Fahrenheit (0 degrees Celsius), on a baking sheet, and put it in the oven. The ice started boiling almost immediately. It was entirely gone in 17.1 seconds.

SpaceX engineer Jon Edwards once calculated the energy being transferred on a hot Florida day into a rocket full of densified propellant. It was about 1 million joules per second, or a megawatt. That's about the same amount of energy consumed by 500 homes running air conditioners on a summer's day. And all of that energy was being pumped into a skinny rocket about 200 feet tall. Now, perhaps, you have some appreciation for the urgency Musk felt about loading his Falcon 9 rocket and then immediately launching.

Muratore served as launch director for the AMOS-6 mission, which carried an Israeli satellite and was due to launch on September 3. Two days prior to the planned launch date, Muratore led the countdown team in conducting a static fire of a new rocket. This was standard procedure for new rockets at SpaceX, and to save two days in the vehicle's preparation, the company took the fateful step of performing the test with the 12,000-pound Israeli satellite attached. Fueling operations proceeded nominally, and the launch team did not have to babysit any abnormal readings. As much of the countdown was now automated, the team mostly watched as the ground systems did their thing—passing through the tens of thousands of sequences, checks, and gates that teams of engineers had preprogrammed.

Everything was fine until eight minutes before engine ignition. Then, things were not fine.

"I saw the first explosion," Muratore said. "It came out of nowhere, and it was really violent. I swear that explosion must have taken an hour. It felt like an hour. But it was only a few seconds. The second stage exploded in this huge ball of fire, and then the payload kind of teetered on top of the transporter erector. And then it took a swan dive off the top rails, dove down, and hit the ground. And then it exploded."

It was 9:07 AM local time. For the second time in fifteen months, SpaceX lost a rocket and a payload. This one was far more painful because the company need not have lost the AMOS-6 satellite. They could have added forty-eight hours to the launch campaign, not attaching the payload until after the static fire test. Worse still, SLC-40 was destroyed. Meanwhile their launch pad at Vandenberg was still undergoing upgrades to accommodate the Full Thrust rocket and densified propellants. And the rebuild of NASA's LC-39A was not yet complete.

"That began one of the most difficult times for SpaceX," Muratore said. "We were in an existential crisis. We were a launch company without any launch pads. And we had a rocket that we couldn't fly."

That morning Musk was asleep at home in Los Angeles. Typically, he works late into the night at SpaceX or Tesla, until two or three in the morning. Then he sleeps for a few hours, until after sunrise. That weekend he had been planning to attend the climax of the Burning Man festival in Nevada, so Musk hoped to store up a few extra hours of rest before long nights of revelry in the desert. As Musk slept, some of his key lieutenants gathered around Shotwell's office at the company's headquarters, texting with a group of people who managed Musk's affairs across SpaceX, Tesla, and other interests. Among them was his assistant, Elissa Butterfield.

Someone had to tell the boss about the rocket, but no one really wanted to. Finally, Butterfield called the security detail at Musk's home and had him woken up, just before 7 AM local time. Almost immediately afterward, Butterfield telephoned Musk directly and said she was connecting him with the "anomaly call." Musk's response to this was puzzlement. What anomaly? Butterfield placed him immediately into the call. It was Shotwell who had to break the news about the accident.

Shotwell had come into the office early that morning, as was her habit. She was the steady hand on the ship, helping calm nerves among employees reeling from a second failure. Shotwell immediately started calling the rocket's customers, trying to reach them before they saw the news. She helped craft a public statement issued about ninety minutes after the accident, acknowledging an "anomaly on the pad resulting in the loss of the vehicle and its payload."

About the same time the statement was released, Musk arrived at the office, grumbling about never being able to find time to sleep. He immediately asked what had happened, but no one had any good answers.

The SpaceX leadership team's mood was not improved later that morning after reading a widely shared Facebook post written by Mark Zuckerberg, the founder of the social network. Zuckerberg had leased part of the bandwidth on the AMOS-6 satellite for some areas of Africa to have internet access to Facebook. "As I'm here in Africa, I'm deeply disappointed to hear that SpaceX's launch failure destroyed our satellite that would have provided connectivity to so many entrepreneurs and everyone else across the continent," Zuckerberg wrote.

Shotwell's response to this was blunt: "What an asshole." The sentiment was widely shared.

## When the smartest people in the industry have no answers

When a fireball engulfed the Falcon 9 rocket that morning, engineers in the Florida control room gasped. Some screamed in shock and surprise at the violent explosion that had come from nowhere. An operations engineer responsible for coordinating with the Range, Julia Black, was among the quickest to recover from the momentary trauma and bewilderment.

She telephoned the 45th Space Wing and managed to get the range to scramble a helicopter to fly overhead and begin assessing damage.

As Muratore started making phone calls to Zach Dunn and others to explain what had happened, Ricky Lim drove over to the perimeter of the launch pad, where he met with first responders. There, a huge black and orange inferno burned for more than two hours before dying down. Photos taken by the helicopter showed the magnitude of damage to the rocket and its launch pad. The only substantial piece of the Falcon 9 that remained intact was the large metal "Octaweb," the structure that supports the nine Merlin engines at the base of the vehicle. Part of the first stage kerosene tank also remained, with the fuel burning off.

"It looked like a cauldron, this awful bowl of smoldering liquid," Lim said.

As usual, Hans Koenigsmann led the failure analysis. He had only completed his last failure investigation eight months earlier. Now, the entire company looked to him, once again, to solve another rocket mystery. Robb Kulin, who was launch chief engineer for the mission and responsible for its safe and successful flight, was already at the Cape, where he had been on console during the explosion. He oversaw the investigation of the pad. The investigators needed clues, and the best ones in Florida were pieces of the rocket. No fragment was too small, because no one knew which bit of evidence might unlock the cause of the accident. The goal was to find the pieces of debris nearest the point of explosion, to suggest its origin. Engineers, technicians, and good ole Florida boys with airboats ventured into the swampland surrounding the launch complex at the northern end of Cape Canaveral. Overnight, the company's safety briefings went from warning about hazardous fuels to cautions about water moccasins.

SpaceX spared no expense in this search. Musk even approved spending half a million dollars to enlist a NASA aircraft carrying a special instrument, known as the Airborne Visible/Infrared Imaging Spectrometer (AVIRIS), to scan the marshland to identify debris by differentiating the composition of materials there. By then the teams were a couple of weeks into the investigation, and Kulin did not believe using an infrared scanner would add much value since it had fairly low resolution and an inability to see through foliage. He emailed Musk and suggested they call off the NASA jet and save the money. This earned him a terse phone call from Musk saying the company needed to be doing "every fucking thing" it could to find debris.

At the time, Kulin was in the final stages of interviewing to become an astronaut, and NASA allowed him to reschedule at a later date while he worked on the investigation. (Kulin was selected as an astronaut in 2017 but left the program a year later for personal reasons.) The NASA plane, which was mobilized quickly, ultimately did fly over the Cape for SpaceX. It turned up no new critical debris. The search continued for many weeks because material had been blasted far and wide. The most important bits near the source of the explosion had the most energy behind them, so they were ejected farthest and were therefore the most difficult to find. As the search area widened, a Category 5 hurricane developed in the Caribbean Sea and, for a time, appeared poised to strike Cape Canaveral. Although Hurricane Matthew turned and narrowly missed directly striking Florida, the threat of the storm barreling ashore and carrying away critical evidence further heightened the urgency of the search.

As pieces of the rocket were collected, some engineers converted the company's brand-new payload fairing processing facility into a hangar where the debris could be laid out in the manner of an airplane crash investigation. Engineers from SpaceX, NASA, the Air Force, and the National Transportation Safety Board assessed this debris for clues. By identifying where the littlest pieces of debris were concentrated on the rocket, the inspectors understood where the most energetic explosion occurred. Similarly, they sought to diagnose how the explosion had propagated from burn marks on the hardware.

It was a desperate time. Kulin's day began at 6 AM, and he was going full throttle two hours later after four coffees. The search and investigative work proceeded until about midnight, after which Kulin and some of the other engineers would meet to unwind from the day's stress until 1 or 2 in the morning. The next day the whole process started over and lasted for weeks. The work environment near the pad was hazardous, with sweltering heat and burnt carbon dust everywhere. "The emotional and physical toll were absolutely terrible, both from missing sleep and the toxic materials all over," Kulin said. It was a raw and relentless experience for all involved.

Koenigsmann and Kulin were desperate for any insights. The problem was not a lack of data. Because the failure occurred on the ground, the company had a trove of data streaming back from sensors all over the first and second stages of the rocket, more than 3,000 channels of video and telemetry. The issue was determining which pieces of data in this haystack were useful. It had happened frighteningly fast. Just ninety-four milliseconds—less than one-tenth of a second—passed between the first bit of anomalous data and the destruction of the second stage.

The night before the static fire test, Koenigsmann had placed his mobile phone next to his bed before nodding off to sleep at his home in San Pedro, California. When he woke the next morning, he saw a text from Kulin.

"Vehicle just blew up on the pad," Kulin texted just four minutes after the explosion.

They talked by phone for a few minutes, discussing the immediate aftermath of the explosion. As a senior manager of flight reliability, Kulin was on Koenigsmann's team.

About an hour later, Kulin texted again. "Looking more like a cryo copv in lox tank again."

Why?" Koenigsmann asked.

"Cryo pressure drop beforehand, and rapid tank pressure rise. Fuel was already full, as was fuel tank COPVs."

In other words, with the second stage tank full of densified oxygen, one of the three helium bottles inside showed a drop in its pressure, followed by a rapid rise in the pressure of the oxygen tank. And then, *boom*. Almost from the outset, engineers identified an issue with the helium COPVs inside the second stage. This was eerily similar to the failure fifteen months earlier, when a liberated pressure vessel had ruptured the second stage oxygen tank. But what, exactly, had caused the pressure to drop inside the helium bottle? That question would bedevil Koenigsmann and his investigation team for more than a month.

These helium bottles are not particularly small. Each one stands about five feet tall and two feet in diameter, about the size of a bathtub. Three of the bottles are bolted to the interior of the rocket's second stage. As liquid oxygen flows into the second stage during the fueling process, the helium COPVs are submerged. They remain so until the second stage separates from the first stage and ignites its Merlin engine. Then, helium is released during the second stage's flight. As liquid oxygen drains to the engines, helium fills the empty space left behind to maintain pressure.

For SpaceX, the best way to lock down the cause of the explosion was replicating the failure of the helium bottle while it was submerged in liquid oxygen. The company attacked this in a brute force manner, with messy testing in McGregor, where engineers and technicians destroyed dozens of pressure vessels in different ways.

"AMOS-6 was so bad," Koenigsmann said. "We blew up thirty COPVs after that. Each COPV had the explosive power of four pounds of TNT. So every test basically disassembled the test stand. We would build test stands and then just blow them up one after the other, trying to figure out what happened. It was like a true mystery."

The frequent meetings of the accident investigation team pulled in senior leaders of SpaceX, including Koenigsmann, but also key figures such as Mark Juncosa and Dunn. They were the smartest engineers at SpaceX, and some of the smartest in the entire industry. And they were genuinely puzzled by the cause of the explosion.

They had no answers for weeks on end.

## Smoking gun

In those early weeks of the investigation, the company's engineers were asked to run down thousands of possible scenarios by which the pressure vessel inside the liquid oxygen tank could have ruptured. This included some pretty wild possibilities. Video of the accident showed a flash on the roof of a building about one mile from the launch site. Leased by United Launch Alliance, the building is now known as the Spaceflight Processing Operations Center.

Could there have been a sharpshooter on the roof?

Musk took to this idea immediately, already considering it a possibility just hours after the accident occurred. The explosion was maddening because it came at a moment when nothing was moving on the rocket but propellant. The countdown was perfect, and engine ignition remained minutes away. A rocket in this state should not explode. It's like a car in the driveway with a half-tank of gasoline suddenly combusting. So the idea of an outside actor appealed to Musk, who indulges in conspiracy theories.

About a week after the accident, Musk alluded to the sabotage theory on Twitter. During the early morning hours of September 9, Musk tweeted, "Particularly trying to understand the quieter bang sound a few seconds before the fireball goes off. May come from rocket or something else." Despite this mention of a "quieter bang" and "something else," the sniper theory remained internal for a while longer.

Some circumstantial evidence did support the theory. Data from onboard the rocket pinpointed the first signs of a rupture about 200 feet above the ground, on the side of the vehicle facing southwest, toward the ULA building. Additionally, the timing of the flash—caught on just a few pixels of video—matched the interval it would take for a projectile to travel from the building to the rocket. Even so, the idea of a sniper stretched the credibility of most SpaceX engineers. That the company considered it underscores the desperation to find a root cause of the accident and the layer of paranoia that began to envelop the investigation. In those first weeks there just was no simple explanation. The physics of a pressure vessel failure were subtle, extremely complex, and elusive. A sniper with a gun offered a straightforward explanation.

It fell to Ricky Lim, the Florida site director, to make inquiries of United Launch Alliance. He did not want to. Lim knew that in making a request to go and look at the building where the flash had been seen, he would sound crazy. In his request, Lim tried to explain that SpaceX was simply performing its due diligence and chasing down every probability out of more than 10,000 potential causes. But it was embarrassing for him, personally. And understandably, ULA officials refused. It did not help when, two weeks later, SpaceX's implication of sabotage was leaked to the *Washington Post*.

After this, the highly improbable "ULA sniper" theory burst into public view. Fortunately for Lim, he was not named in the *Post* or subsequent articles. But it still stung his pride, and that of the company, to be ridiculed for being seen to blame a competitor when the catastrophic failure of the AMOS-6 mission was certainly SpaceX's fault.

It particularly hurt to show weakness when it came to United Launch Alliance. By mid-2016, SpaceX and ULA had been rivals for more than a decade. For much of that period it was a lopsided rivalry. ULA launched the nation's most important national security and science missions, and SpaceX was a pesky gnat. But after years of playing a distant second fiddle to United Launch Alliance, SpaceX had started to catch up. Even with the AMOS-6 failure in 2016, the Falcon 9 would tie ULA's Atlas V rocket for the number of launches by an American rocket that year, at eight. It was the first time SpaceX matched its competitor.

This was a searing hot competition. While public statements from both companies are often papered over with a thin veneer of congeniality, the rivalry between SpaceX and ULA is a bitter one. Back in 2005, years before SpaceX reached orbit, Musk sued in U.S. District Court to stop the merger of Boeing and Lockheed Martin's rocket businesses into a single company, United Launch Alliance. He cited antitrust reasons.

Two years later, of course, ULA and its parent companies returned the favor by lobbying heavily to prevent SpaceX from acquiring a launch site at Cape Canaveral. United Launch Alliance had also happily joined forces with Blue Origin to try and prevent SpaceX from gaining sole control of Launch Complex 39A at Kennedy Space Center.

One of the most acidic battles played out in 2014 and 2015, over access to national security launch contracts. In April 2014 Musk took the exceptional step of suing the U.S. Air Force for its decision to award three dozen launches to ULA, alone, for military missions through the remainder of the decade. The "block buy" contract was valued at $11 billion, a huge sum of money. At the time, the Falcon 9 rocket had flown a total of nine missions, all successful. What SpaceX wanted, Musk said, was an opportunity to compete for some of these launch contracts. In return, he offered a substantially lower price.

"The ULA rockets are basically about four times more expensive than ours, so this contract is costing the U.S. taxpayers billions of dollars for no reason," Musk said at the time.

Some military officials thought SpaceX's concerns were justified. Wayne Monteith, the Range commander who approved landing the first Falcon 9 back at Cape Canaveral, then served as a senior military advisor to Secretary of the Air Force Deborah James.

"It was a shame that Elon had to do it, but quite frankly he had to do it," Monteith said. "Because the Department of Defense was really not set up at that time to embrace innovators with an entrepreneurial mindset."

SpaceX and the Air Force resolved their claims nine months later. As part of the sealed settlement, the Air Force agreed to open up some of its national security missions in the 2010s to competitive bids and promised to expedite certification of the Falcon 9 rocket for its most valuable payloads. Amid this dispute, the chief executive of ULA, Mike Gass, was fired and replaced with a more dynamic leader, Tory Bruno, who drew leadership inspiration from the Knights Templar. Bruno had an unenviable task. Over its first decade, United Launch Alliance had grown fat and complacent on its monopoly of military launch contracts. And the company was seriously hamstrung by its two owners, Boeing and Lockheed, which had taken profits from the rocket-maker rather than investing in it.

Balding and gallant, Bruno is generally well liked in the aerospace community. Under his leadership, ULA has slashed its executive ranks, cut its overall workforce, and reduced the number of pads it launches from in an effort to become more cost competitive. ULA has even taken tentative steps toward rocket reuse but has yet to perform any in-flight experiments. Both ULA and SpaceX launch from Vandenberg Air Force Base in California, and Cape Canaveral in Florida. During an interview,

I asked Bruno whether SpaceX was a good neighbor at those launch sites. At this question, Bruno laughed. And then he responded, "They definitely are our neighbors." In other words, no.

SpaceX engineers even tested the ULA sniper theory in McGregor, shooting at COPV tanks with a rifle to determine whether they would explode, and what the resulting detritus looked like. The results were inconclusive. Ultimately the sniper theory investigation was dropped when the Federal Aviation Administration sent SpaceX a letter saying there was no gunman involved, period.

The real breakthrough came in McGregor later that fall, with repeated pressurization tests of a liquid oxygen tank with COPVs inside. The test team in Texas had gotten a COPV to auto-ignite while immersed in liquid oxygen. A camera inside the tank had recorded it all. This was the smoking gun.

This long-awaited recreation of the failure brought a huge sigh of relief to Koenigsmann, the accident investigation team, and the entire company. It allowed SpaceX to begin planning a fix and a return to flight. What the engineering team had not appreciated was the delicate physics playing out inside the pressure vessels. A COPV consisted of an inner aluminum liner, with carbon fiber wrapped around the outside for strength. For the sake of speed, the test and launch teams filled the COPVs with helium as quickly as the ground systems would allow. This made the aluminum liner very hot. Done fast enough, this could turn the pressurized vessel from a flexible carbon fiber into something that more closely resembled brittle glass, allowing a "buckle" to form in the liner. During propellant loading, super-chilled oxygen could pool in these buckles. The combination of heat and pressure solidified this trapped oxygen, which was pushed so hard into the carbon fiber it would ignite.

SpaceX had successfully launched eight missions using densified oxygen before the AMOS-6 failure. For each of those missions Lim,

Muratore, and other launch engineers had gradually updated the count-down to lift off with the coldest possible liquid oxygen onboard the rocket. Simply, they had been working toward Musk's goal of loading faster and faster. During the AMOS-6 static fire test, the Falcon 9 rocket hit an unlucky combination of propellant loading timing and temperatures to compress the liquid oxygen and ignite it.

For a time, the launch team *had* been fucking flawless. But in their efforts to rapidly load propellant, running that desperate race to keep super-chilled propellants on the rocket as cold as possible, they had gone too fast. The fix was straightforward. Over time, the helium bottles would be redesigned. In the near term, SpaceX reverted to an earlier, slightly slower, and proven method of fueling the rocket.

## Everything could come crashing down or the day could be amazing

A line of SpaceX-chartered buses waited in the dark, outside the Ritz-Carlton Bacara resort in Santa Barbara, during the predawn hours of January 14, 2017. Among those mingling outside the seaside hotel was Matt Desch, the Iridium chief executive. He was nervous as hell. Eight years had elapsed since Desch closed the deal to fly his Iridium satellites on the Falcon 9 rocket, and now that day had come. There were only two outcomes. His grand plans could come crashing down into the Pacific Ocean, or the day would be amazing. Desch feared the worst.

Iridium originally planned to launch its first two new satellites on a smaller, stubby rocket based on an old Soviet intercontinental ballistic missile. Russia started to sell this Dnepr rocket commercially at the turn of the century to Western customers to loft small- to medium-sized satellites. In late 2001 and early 2002, Musk had traveled to Russia three times to buy a launch on the Dnepr to fly a small greenhouse to Mars.

After each trip he returned empty-handed, because Russian negotiators would not provide a firm launch cost. "After that last trip from Russia, I was like, man, the price just keeps going up," Musk said. His frustration, in part, impelled Musk to found SpaceX.

Desch had been able to come to an accommodation with Russia for the rocket, which sells for about $30 million, or half the price of a Falcon 9. He liked the idea of flying on a smaller booster first, because then he would put just two Iridium satellites at risk. If there was a problem with the satellites in space, the rest of them could be fixed on the ground. Although the Falcon 9 would still launch the majority of the constellation, Dnepr could be used for spares.

But in the spring of 2014 Russia invaded the Crimean Peninsula, territory held by Ukraine. This limited access to the Dnepr rocket, which was partially manufactured in Ukraine. To see what could be done, Desch met with the Russian Ambassador to the United States, Sergey Kislyak. Iridium was ready to launch, and Kislyak assured Desch there were no problems. He would see about getting launch approval from the Russian military.

Several weeks later, Kislyak admitted that the Russians were halting the launch of Dnepr for commercial customers. He offered a Soyuz rocket as an alternative. This was very large for just two satellites and would cost significantly more. So Desch went back to SpaceX, which had been steadily increasing the performance of its Falcon 9 rocket through the 1.1 and Full Thrust upgrades. Could the company carry ten satellites on each of its rockets instead of nine, he asked? SpaceX could.

This was all well and good, but it meant that Iridium would not get its test launch. Instead of risking $50 million worth of spacecraft, Desch had a quarter of a billion dollars' worth of hardware riding on the Falcon 9. And then the AMOS-6 failure happened.

"When that thing blew up on the pad in Florida, we were supposed to launch a month and a half later," Desch said. "We were next in the queue."

Iridium had little choice but to take the return-to-flight mission. Iridium's original satellites were nearly two decades old, and there were almost no spares left in orbit. With the deal for a Dnepr launch falling through and delays to the Falcon 9 rocket after its successive accident, the second-generation Iridium NEXT replacement constellation of satellites was years late in getting to space. Changing rockets also was not a realistic option. Iridium would have had to pay twice the price and taken a year or two to adapt to a different launch vehicle. It was the Falcon 9 or bust.

On the morning of January 14, therefore, it is not really too much of an exaggeration to say that the fate of two companies rode on that new Falcon 9 rocket. SpaceX had to restore confidence in its workhorse vehicle. And Iridium could ill afford the financial loss of ten satellites and potential for service outages. It might not be a death blow for the $3 billion Iridium NEXT deal Desch had cobbled together, but it might be close. Investors of the publicly traded company would be furious, and the plan's bankers, insurers, and satellite manufacturers would all be howling to be paid back.

"If that launch was unsuccessful, the next day was going to be really shitty," Desch said. "And maybe the next week, and maybe I would have gone off and done something else. I'd never been so stressed out in my life."

As the buses rolled out of Santa Barbara, Desch stared down at his phone, a bright beacon in the dark morning. Text updates were pouring in from the company's chief operating officer, Scott Smith, inside Iridium's control room. Brisk winds the night before had blown down the cable carrying communications and power from the launch tower to

Iridium's satellite. For the time being, Iridium could not communicate with its satellites to know they were healthy. To further Desch's agitation, the bus driver got lost on the winding roads of the sprawling base, which spans 100,000 acres along California's Pacific Coast. After the bus veered left and wiped out a stop sign, Vandenberg police officers pulled it over and held a heated discussion with the driver outside for fifteen minutes. Finally, the officers escorted the bus to the Vandenberg viewing site.

By this time the skies were lightening. As the Iridium employees and their guests milled beneath a white tent, eating croissants and drinking coffee, Desch continued to check his phone. A SpaceX employee was in a JLG lift, attempting to reconnect the downed communications cable before fueling of the rocket could commence. Meanwhile, there was a range "violation," meaning a boat or plane had moved temporarily into the protected area around the launch site. And winds higher in the atmosphere, which can buffet a rocket and break it apart, were nearing unacceptable limits. Amid all this uncertainty, Desch stood before an audience of about 150 people and gave a short speech. In truth, these distractions helped Desch. They prevented him from focusing on the fact that for him and his company, the all-or-nothing moment was nearly at hand.

About twenty minutes before the launch, SpaceX began its webcast. For most of the company's early missions, a seasoned engineer named John Insprucker provided matter-of-fact commentary about the launches. However, in mid-2015, Musk was convinced to try a new approach, as a means of boosting the company's ability to recruit engineers.

Musk had always been lukewarm on the idea of a webcast. He wondered aloud about what purpose it served, noting that airports did not have webcasts for airplane takeoffs. If SpaceX were aiming to make rocket launches routine, should it make a big deal about webcasts? Upon realizing that customers liked them because they could advertise their services to the hundreds of thousands of people who watched, Musk assented. He also came to understand how these shows helped build the company's fan base.

The new webcast format with a multitude of hosts, often speaking from the energized floor of the factory amid hundreds of excited young engineers, took this appeal to a new level. SpaceX held a companywide audition in 2015. Rehearsing for webcasts often occurred during the daytime, and it was extra work an employee had to take on. But it also offered a platform for younger engineers to increase their visibility and for SpaceX to burnish its image as a young, hip place to work. Among the first hosts was Lauren Lyons, a Black female in aerospace engineering who brought a new vibe to the SpaceX stream. She injected some levity into what had previously been a fairly staid broadcast.

"Each of the ten spacecraft that are onboard right now have a mass of 600 kilograms," she said of the Iridium satellites. "And when the solar arrays are fully deployed they have a thirty-foot wingspan, which equates to more or less four Shaquille O'Neals."

At this point the rocket entered the final countdown. The range had cleared. And though marginal, the winds were acceptable. At Vandenberg, Desch and his team looked across rolling hills to the launch pad. In the final moments before liftoff, Kathy Morgan, Iridium's chief legal officer, and Bryan Hartin, the company's head of sales, had quietly stepped up beside Desch. Sensing the stress overwhelming their boss, they had taken those positions to catch him if he fainted or otherwise buckled in the tense moment.

The nine Merlin engines lit, and the Falcon 9 rocket ascended. But from the vantage point of those assembled at the viewing site, the rocket did not climb very far. "We're watching it go up, and then it's just hanging in the air," Desch said. "It looked like it was just not going to make it."

But he did not faint. And the rocket did not tumble back to Earth. What Desch did not realize until afterward is that the viewing site was north of the launch pad at Vandenberg, and the rocket was traveling downrange, away from him, toward the horizon. As it pitched over, assuming a more horizontal posture to enter orbit, the illusion of a rocket going nowhere fast was enhanced. But everything was actually just fine. As the callouts from Lyons and others on SpaceX's webcast continued to express confidence in the flight, a huge wave of relief washed over Desch and the other guests.

Yet for Iridium and SpaceX the mission remained far from over. The satellite dispenser, connected to the rocket's second stage, still had to release its cargo at the proper time. Then the satellites had to wake up and communicate with Iridium controllers back on Earth. The entire process would take another hour and a half. So Desch left the viewing party with a couple of his board members and went up to SpaceX's launch control center. After he arrived, Iridium began to get "heartbeats" from each of the satellites as they passed over a ground station on Svalbard, an archipelago of islands between the North Pole and Norway.

Musk did not attend every launch, but this flight was critical to the future of his rocket company. He had been nervous about the flight, too, staying up all night. Inside the SpaceX control center, he stood cloistered with some of his friends from his days with the PayPal venture, a group informally known as the "PayPal Mafia." Desch asked someone with SpaceX whether Musk might be willing to come by the post-launch celebration that afternoon at a nearby winery. He was told that Iridium was SpaceX's most important commercial customer, and Musk would likely do almost anything at the moment.

When Desch asked Musk, his reply was simply, "Now?" As in, he was ready to depart right at that moment, if needed.

Musk did come to the party later in the day. He wore his typical attire, black jeans and an Occupy Mars T-shirt, and posed for pictures with guests who wanted a memento. As he mingled at the party, Musk was fairly quiet. Some of that was no doubt due to exhaustion. But it was also due to stress. During tense moments the levels of cortisol, the primary stress hormone, spiked in his body. Like Desch, he, too, felt mostly relief that on a day when a million things could have gone wrong, his rocket and its payloads had followed the single, narrow path to a successful outcome.

After a very trying eighteen-month period in which his company had endured two catastrophic accidents, SpaceX was back in space, and back in business.

| 10 |

# THE COST OF MARS

*September 2016*
Guadalajara, Mexico

In the dark days following the loss of the Israeli satellite in September 2016, organizers of the world's largest international space conference reached out to officials at SpaceX. They were sorry to hear about the accident and completely understood Musk's need to postpone his appearance as a keynote speaker that year.

Musk was due to deliver a much-anticipated speech less than four weeks later, in Guadalajara, Mexico, at the International Astronautical Congress. In April, when asked about his vision for settling Mars, Musk had replied, "I'm planning on giving a talk at IAC, which is a good venue to describe that approach. I think it's going to sound pretty crazy, so at least it should be entertaining." Then SpaceX lost its second rocket in sixteen months, and it seemed foolish to spin tales of sending humans to Mars—hardly the time for "crazy" talk.

Musk did not care.

In response to the outreach from the space conference, Musk was asked point blank if he wanted to postpone. The organizers assumed he would, but Musk told them he was still coming to Mexico.

The AMOS-6 accident that summer, and Musk's refusal to let this near-term setback detract from his Mars plans, highlighted one of his defining traits. Musk always kept one eye firmly fixed on the big picture.

"In times of urgent problems, you still need to make time for the important long-term vision," one senior SpaceX official said. "That's how you accomplish big things. You're going to have one-off problems, but giving a seat at the table for the future is why people came to work at SpaceX."

During the ninety-minute speech in a packed auditorium, Musk was at times a dreamer, a comedian, a nerdy engineer. He laid out an exceptionally bold vision. NASA's Apollo space program, which required 400,000 people and 5 percent of the nation's budget, put all of two people at a time on the Moon. Musk proposed a transportation system—he would not settle on the Starship name until a few years later—that would carry 100 people to Mars at a time.

Musk also sought to answer the *why* of going to Mars. Why should SpaceX, why should NASA, why should humanity invest the time, money, and hard work in building a fleet of spaceships and settling Mars? Musk expressed the need to ensure "that the lamp of consciousness is not extinguished." What would happen if humanity failed to settle other worlds? "We're confined to one planet until an extinction event," Musk said. This does not mean abandoning Earth or chewing through its resources. Earth should be protected and preserved, he said. But eventually an asteroid, particularly virulent pandemic, or a nuclear conflict would end our civilization. To avoid this, we must become a multiplanetary species. Mars is far from a paradise. Indeed, it is far less hospitable than the most barren desert or frozen tundra on Earth. But it is the closest, best place to start living off-world.

The most striking thing about the speech is that Musk let it all hang out in the wind. He put his entire vision out there, and it felt like this incredibly vulnerable moment. Enthusiasts had talked like that before about space settlement, but no one had taken them seriously. They were eccentrics or crackpots. But Musk had a real rocket company, and some of the smartest engineers in the world. They'd just landed a rocket on a boat in the middle of the ocean, for goodness's sake. He was credible . . . and putting his credibility on the line.

In my reporting on the speech, I described the plan as "audacious." As for Musk, I said his speech might be "madness, or brilliance—or both." Eight years later, the audacious plan is well afoot, and possibly more brilliant than mad.

## How does one build a city on Mars?

Musk has been speaking of Mars since the earliest days of SpaceX. The first time Gwynne Shotwell met with Musk, in 2002, he spoke passionately about his vision for settling the red planet. "He was talking about Mars, his Mars Oasis project," Shotwell said. "He wanted to do Mars Oasis, because he wanted people to see that life on Mars was doable, and we needed to go there." She thought Musk slightly strange to be sermonizing about Mars when SpaceX had yet to cut a sheet of metal for its first Falcon 1 rocket, but Shotwell harbored no doubts about why Musk had gotten into the launch business.

For the first decade he let SpaceX focus on smaller rockets, but by the mid-2010s Musk started to lay down some preliminary plans. He went to Tom Mueller, then serving in an advisory role as chief technology officer of propulsion, and asked him what it would take to land a hundred tons on the surface of Mars. The most mass ever set down on Mars was a single ton, by NASA's $3 billion Curiosity rover two years earlier, in 2012. This mission's novel "sky crane" landing technique required a sequence of steps so risky that its designers characterized it as "seven minutes of terror."

Now Musk wanted to land one hundred times that amount.

Originally, Mueller and Musk envisioned using liquid hydrogen to fuel the Mars vehicle's engine. Liquid hydrogen is by far the most efficient fuel to burn in rocket engines, especially for moving through space. A spaceship can go farther, using less fuel, with hydrogen compared to other propellants. However, as Mueller delved into his calculations he found that methane, the main constituent of natural gas, might be a better option. Although it lacks the fuel efficiency of hydrogen, liquid methane is easier to work with and costs less to fuel a rocket. Moreover, it is denser than hydrogen, so methane fuel tanks could be significantly smaller. And like hydrogen, methane could be produced on the surface of Mars, allowing rockets to be refueled for a return trip to Earth.

Musk had Paul Wooster, who later became SpaceX's "Principal Mars Development Engineer," check Mueller's numbers. They were sound, and Musk agreed to switch to methane as a fuel for his Mars rocket. "That was a huge decision," Mueller said. "Then the question was, what would it take to get that stage back? And it's actually harder to get back."

That's because SpaceX needed a lot of methane and liquid oxygen to fly home. To bring even a small amount of cargo—and astronauts—back from Mars would require topping off the rocket's tanks with more than 1,000 tons of propellant and oxidizer. This is not a trivial amount. On Earth it takes about fifty full-size tanker trucks to deliver this much fuel. And there are no tanker trucks on Mars. Or gas stations.

The process of producing propellant on Mars is theoretically simple. NASA uses a chemical reaction known as the Sabatier process on the International Space Station to derive water from carbon dioxide and hydrogen. Methane is an unwanted byproduct of the reaction and is jettisoned overboard. The same reaction could be used on Mars, while retaining the methane. The potential supply of fuel is essentially limitless; Mars has plenty of water ice to produce hydrogen, and the planet's thin atmosphere is more than 95 percent carbon dioxide. The problem is not raw materials but the energy needed to produce propellant.

During his Guadalajara speech, Musk addressed the rocket's fuel needs. The company ruled out kerosene, which is burned by Falcon rockets, because it could not be produced on Mars. Liquid hydrogen requires larger storage tanks, as mentioned, and is difficult to work with. It must be stored very near absolute zero or it will boil off. "We think methane is really just better across the board," Musk said. "The trickiest thing really is the energy source, which we think we can do with a large field of solar panels."

A large field, indeed. Mueller and Wooster calculated it would take approximately 750 kilowatts of continuous energy, for two years, to produce 1,000 tons of liquid oxygen and methane. This is a big ask on Mars. Due to its distance from the Sun, the red planet receives less than half the energy from the Sun that Earth does. Therefore, it would take something like an acre of solar panels producing energy for two years to make the requisite fuel to come home from Mars. That is possible, but difficult. Most people flying to Mars on Starship will probably be taking a one-way ride.

Such engineering challenges don't faze Musk. If physics does not prevent something, then by definition it can be done. In early 2020 I drove to the company's rapidly expanding Starship build site in South

Texas where they were building the world's largest rockets beneath giant white tents. Musk was eager to talk about overcoming the challenge of settling Mars. He had been thinking a lot about this, and I wrote about his messianic zeal at the time for *Ars Technica*, where I am the senior space editor. The factory hummed and buzzed with incredible energy as the company began to build massive rockets beneath tents.

"We need to have a self-sustaining city on Mars," Musk said. "That city has to survive if the resupply ships stop coming from Earth for any reason whatsoever. Doesn't matter why. If those resupply ships stop coming, does the city die out or not? In order to make something self-sustaining, you can't be missing anything. You must have all the ingredients. It can't be like, 'Well, this thing is self-sustaining except for this one little thing that we don't have.' That'd be like saying, 'Well, we went on this long sea voyage, and we had everything except vitamin C.' Okay, great. Now you're going to get scurvy and die—and painfully, by the way. It's going to suck. You're going to die slowly and painfully for lack of vitamin C. So we've got to make sure we've got the vitamin C there on Mars. Then it's like, okay, rough order of magnitude, what kind of tonnage do you need to make it self-sustaining? It's probably not less than a million tons."

More than half a decade after Mueller's original calculations, Musk still planned on individual ships bringing 100 tons to Mars. This had been baselined into the design of the Starship rocket. Yet even with this remarkable amount of tonnage per delivery, it would take a mind-boggling 10,000 Starship landings to reach one million tons. Never mind that in half a century of Mars exploration, there have been all of about ten successful landings, with a combined total of a few tons of payload.

Starships are designed for reusability, of course. But due to the difficulty of refueling on Mars, most of them probably weren't coming back to Earth. So Musk had to make his Starships cheap. He said he wanted to build them for $5 million each. That may sound like a tidy sum of money, but even pedestrian spaceships, by comparison to Starship, can literally cost billions of dollars to build. And Starship is not pedestrian. It is the largest and most powerful upper stage of a rocket ever constructed. And Musk wanted to build them for five million bucks.

I couldn't help myself, blurting out: "That's fucking insane."

"Yeah, it's insane," Musk replied.

"I mean, it really is."

"Yeah, it's nuts."

"As I look across the aerospace landscape, nobody is doing anything remotely like this," I said.

"No, it's absolutely mad, I agree," Musk said. "The conventional space paradigms do not apply to what we're doing here. We're trying to build a massive fleet to make Mars habitable, to make life multiplanetary. And each of those ships would have more payload than the Saturn V—and be reusable."

The challenge of building thousands of Starships, and the difficulty of refueling them on Mars, reveals something fundamental about how Musk and SpaceX see their path to the stars. NASA explores. During Apollo, it sent a pair of astronauts on brief sorties down to the lunar surface. For Mars, too, it envisions roundtrip missions for a handful of astronauts, complete with triumphant ticker tape parades when they return to Earth. But Musk does not want to explore Mars. He wants to settle Mars. And that is really a massive difference compared to what humans have done in space for our six decades among the stars.

This is a preposterously colossal undertaking. And setting aside all of the engineering challenges, I was left with one big question.

Who the hell was going to pay for this?

## How does one *pay for* a city on Mars?

Bulent Altan was among the first to know the answer. Born in Turkey, he came to the United States in 2002 for graduate school at Stanford University. Altan joined SpaceX two years later to work on avionics, playing a pivotal role during the Falcon 1 years on Kwajalein. During all of that, and later through development of the Falcon 9 and Dragon, Altan gave his all to SpaceX.

There were downsides. While working feverishly to complete preparations for the Falcon 9's debut, Altan received a call from his sister back in Turkey. Their mom was sick. She was forgetful and having difficulty walking. Could he come home? After the first launch, Altan flew back to Turkey for just the second time in a decade. His mom had been an engineer, helping to inspire Altan's career choice. They had been close, but now he found he could not share the success of the Falcon 9 rocket with her. She just could not follow him. Later, Altan's mother was diagnosed with Creutzfeldt-Jakob disease, a degenerative brain disorder.

"It was a death sentence," he said. "After that she just deteriorated faster."

In October 2012, Dragon flew its first operational mission to the space station. After the launch, Altan went to dinner with some of his closest friends at the company, David Giger and Paul Forquera. While there, Altan received another phone call from Turkey. His mom had died. According to Muslim tradition, a person should be buried as soon as possible. Altan, however, asked his father to wait an extra day because he needed to be present for the spacecraft's capture and berthing. A little more than a day later, when he entered the Mission Ops room, John Couluris nodded toward Altan with an understanding look that said, "I am sorry what happened, but also very grateful for you to be here today."

After Dragon's capture, Altan walked to the back of the room, hugged Couluris, and drove to the airport.

The loss of his mother and the realization of the familial sacrifices he had made hit Altan hard. He returned from the funeral with a new perspective on the all-consuming SpaceX lifestyle. For a time, he soldiered on. He became vice president of avionics, taking responsibility for upgrades to version 1.1 of the Falcon 9. As this booster neared its first launch in the summer of 2013, Altan moved into a house with Tim Buzza and Hans Koenigsmann near the Vandenberg launch site. He witnessed the anguish experienced by Buzza, firsthand, during the Great LOX Boil-Off and in subsequent interactions with Musk.

After his mother's death, Altan did not take this well. The day that Buzza internalized his decision to leave SpaceX, Altan reached a similar conclusion. "Tim was a role model," Altan said. "He probably was my biggest role model in all of SpaceX. When that happened to him, I checked out, too. I had done what I wanted to do. It was time for the next challenge."

To gain some distance from SpaceX, Altan moved to Europe in 2014, where he could still work in the aerospace industry while being a short flight from his family in Turkey. To see what life was like on the other side of the fence—that is, at a large, traditional defense contractor as opposed to an innovate-or-die environment like SpaceX—Altan took a job at Airbus Defense and Space near Munich, Germany.

So what was it like going from a scrappy startup to a conglomerate with 40,000 employees in three dozen countries? Terrible, mostly. The company had fantastic leadership, Altan said, but they could not change the company's culture. It was such a huge difference from SpaceX, where Musk would decide and employees either implemented it, quit, or were eventually fired. By contrast, layers of management at traditional space companies held too much power.

After several months at Airbus, Altan's "head of digital transformation and innovation" title began to feel like mockery. So he emailed Musk in the summer of 2016. Altan had left SpaceX on good terms, and now he said he wanted back into a fast-paced environment. Musk told him the company would find a job to suit his talents.

Musk said Altan could work on one of two new projects: Starship or the company's nascent Starlink satellite network. Given Altan's background in avionics and electronics, the challenge of developing and flying thousands of Starlink satellites appealed to him. This was Musk's preference, too, as the company's early efforts to develop Starlink were struggling. "Elon talked about how the history of large satellite constellations is only full of failures," Altan said. "I could feel his concern about how Starlink was going."

In 2016 high-speed internet from satellites still seemed like an idea from the future. Receiving data from space was not new. By then, DirecTV and DISH Network had been providing television service from outer space for two decades. However, this connectivity came from large, expensive satellites in geostationary orbit. At an altitude of more than 22,000 miles above Earth, the rate at which these satellites orbit the planet exactly matches the speed at which the planet rotates on its axis. This is great, because it means the satellite is always at the same location in the sky.

The problem is that data traveling across tens of thousands of miles to and from these satellites, even at the speed of light, experiences lag. This makes delivering high-speed internet without latency almost impossible. The solution is putting satellites in low-Earth orbit. However, these satellites orbit much more rapidly than the planet rotates, so they are always zipping overhead—the time from horizon to horizon is a matter of minutes. To make broadband internet from space work, there must be a lot of satellites, and they must be very, very smart about communicating with one another so that a user's connection is not lost as a satellite passes over the horizon.

Musk estimated that SpaceX needed about 12,000 Starlink satellites for global internet coverage. That is a mind-boggling number of satellites. No company, or country, in the world operates more than a few hundred satellites. And Starlinks were not particularly small; each was about the size of a kitchen table, with a mass of more than 500 pounds. Developing all of this sophisticated technology, and launching it into space, was probably a $10 billion project at least.

So why would Musk undertake this? SpaceX was already struggling under the weight of his expectations. Altan returned to the company just after the AMOS-6 failure and ahead of Musk's Starship presentation in Mexico. SpaceX had its hands full with returning the Falcon 9 to flight and designing Starship. Now they were supposed to become the world's largest satellite operator, by a factor of more than ten?

"Elon made it clear that Starlink had to work," Altan said. "It was going to be the moneymaker for SpaceX, and fund everything we were going to do in the long term."

In other words, Starlink was created to compete with Comcast and other internet providers, as well as provide a secure communications backbone in remote parts of the world that the U.S. military and its allies might find valuable. This was potentially worth tens, if not hundreds, of billions of dollars. And instead of spreading that value to shareholders, Musk intended to build a fleet of rockets and ship one million tons to Mars to make a self-sustaining settlement there.

During a meeting with his Starlink team about marketing the service to potential buyers, which I sat in on, Musk put this bluntly. "We should remind our customers that they're helping life as we know it become multiplanetary," Musk said. "You could be giving money to fucking Comcast. Or you could help humanity get beyond Earth. Those are your two options."

To pay for Mars, then, Musk made a huge bet in the mid-2010s. SpaceX initiated not one but two massive projects, Starship and Starlink, simultaneously. Neither project had any precedent in history. Starship would be bigger and more powerful than NASA's mighty Saturn V rocket and needed to be reusable many times over. Starlink was far larger and more ambitious than any satellite constellation in history. Both projects were likely to fail.

"Here's how Elon thinks about this," Altan explained. "If SpaceX is an amazingly successful financial company, but doesn't make it to Mars, then SpaceX has failed. And I love that about him. All the theatrics about the person aside, he had that fantastic focus on the final mission. For him, *not* doing these two things is failure, even if in everyone's books SpaceX was already a success."

When he spoke in Guadalajara, Musk did not reference Starlink. Instead, on a slide titled, "Funding," he lightheartedly listed some of the ways that SpaceX might pay for its grand Mars expeditions. These sources included stealing underpants, launching satellites, sending cargo and astronauts to the space station, and a Kickstarter.

"Obviously it's going to be a challenge to fund this whole endeavor," he acknowledged. "We do expect to generate pretty decent net cash flow from launching lots of satellites and transporting cargo to and from the space station. And I know there's a lot of people in the private sector who are interested in helping fund a base on Mars. Then perhaps there will be interest on the government sector side. Ultimately this is going to be a huge public-private partnership."

One reason Musk may not have spoken publicly about Starlink in Mexico was due to the program's struggles. Following Altan's departure in 2014, Musk hired a longtime Microsoft engineer named Rajeev Badyal to help run avionics for SpaceX. Badyal had achieved some successes at Microsoft, such as the Xbox, as well as some failures, like the Zune portable music player. After coming to SpaceX, Badyal took leadership of the Starlink program. To tap into his network of contacts in Washington, near Microsoft, Badyal convinced Musk to move the Starlink offices to Redmond. It was the first product design office SpaceX opened outside of Hawthorne.

Located more than 1,000 miles away and staffed with people from more conservative work environments like Microsoft, the Redmond office developed its own culture. Musk did not approve, and he wanted better oversight of what was happening in the Washington office. As one of the original SpaceXers, Altan had come of age during the scrappy era. Musk dispatched him to Washington as a vice president in charge of satellite mission assurance.

This experiment did not go particularly well. The Redmond office, which had a few hundred employees, or about 5 percent of SpaceX's total workforce, viewed Altan as the spy from Hawthorne. They sought to fly outside the orbit of Musk, doing their own thing. At times Altan would be sitting at his desk and see various teams exiting from meetings. He would ask what the gathering was about. Often, it was a topic like orbital debris.

"The mission assurance guy should also be working on debris," he said. "But I was always an outsider."

The situation felt untenable, and Altan left SpaceX a second time in September 2017. A few months later, in early 2018, SpaceX launched its first two prototype Starlink satellites, nicknamed Tintin A and Tintin B. They worked well enough. But that spring, Musk clashed with Badyal over the pace of development. The former Microsoft engineer wanted to continue tinkering with the Starlink satellite design and fly more prototypes before launching large batches of operational satellites. Musk sent Mark Juncosa to Redmond to investigate, and he confirmed Musk's suspicion that Badyal's office was moving needlessly slow and had burdened itself with requirements.

That was enough for Musk. In June 2018 he flew to Washington, firing Badyal and four other leaders of the Starlink program. He installed Juncosa to right the ship and foster SpaceX's move-fast culture. Less than a year later, SpaceX had its first batch of sixty Starlink satellites on the launch pad. It was a remarkable achievement, both in terms of rapid mass production of satellites previously unseen in the space industry and the sophisticated technology needed for the satellites to relay signals to the ground and among each other. As ever, Juncosa was Musk's essential fixer.

On the eve of the launch, Musk told reporters that the development required to reach this point had been extremely difficult. "There is a lot of new technology, so it's possible that some of these satellites may not work," he said. "There's a small possibility that all of these satellites will not work."

But most of them did. By the end of 2023 nearly 6,000 Starlink satellites were flying in space thanks to the reusable Falcon 9 rocket. After starting as a rocket company, SpaceX now also claimed the mantle of the world's largest satellite operator. Starlink accounted for more than two-thirds of all active satellites in orbit. And critically for the Mars vision, the program had begun to generate positive cash flow.

To reach Mars, SpaceX may not have to steal underpants after all.

## What it's really like to work for Musk

When Musk sought to instill the SpaceX culture in Redmond, he meant pushing forward as hard and as quickly as possible. He expected long hours and excellent work. Lieutenants, like Juncosa, understood Musk's demands and implemented them.

This management style has clear benefits. SpaceX builds some of the very best products in the space industry, faster and at far lower prices than its competitors. The downside is also very real. The environment took a toll on its employees. Frustration with these long hours prompted some directors, like Brian Mosdell in Florida, to quit. He did not feel comfortable enabling this culture.

"I don't know what they think, but eighty hours a week is just completely unreasonable, let alone one hundred hours a week," Mosdell said. "And I don't think they had a genuine sensitivity to that issue. It was more of an excuse to just keep driving. And I think the management strategy was to drive people into the ground, because you can replace them later, and you'll get the biggest bang for your buck."

There were essentially two layers of employees at SpaceX: those who met with Musk, and those who worked for the managers who did. Each group faced its own difficulties. For the managers, there was the challenge of Musk himself. For the lower-level employees, there was the challenge of carrying out his mandates, the incredibly aggressive schedules and seemingly impossible engineering problems. Nearly everyone I interviewed for this book, sometimes on the record and sometimes off, said they worked the hardest they ever worked in their lives at SpaceX. You were either all in, or pretty quickly you were out.

To better understand this culture, it is helpful to consider the experiences of a few employees who loved their time at SpaceX but ultimately turned away.

Robert Rose, the video game programmer who felt awed during his first meeting with Musk, would rise from an entry-level position as a software engineer to become director of flight software in 2011, reporting directly to Musk. Rose came to feel as though he were a character in *Wolf Hall* or *The Tudors*, seeing himself as a Thomas Cromwell–like figure who had risen from humble beginnings into a place at the court of King Henry VIII.

"I recognized that there would come a day when I would have a conversation with Elon that would be my last," Rose said. "I approached every meeting with him as if it would be my last. Basically, I would think, *What do I need to present in this meeting so that I don't get fired?* I was able to separate the emotion from that. A lot of people did not, and they were not able to adopt a healthy mindset for working with Elon."

Rose stuck it out for more than three years in this senior role. During that time, he experienced the kinds of highs that draw people to SpaceX despite the long hours and demanding work environment. He had helped get the Falcon 9 flying and programmed for cargo and crew Dragons. His team had helped write the code that would land Falcon rockets one day. SpaceX attracted employees like Rose because they truly believed in the mission. They could go elsewhere and make more money. But at SpaceX they could quite literally help extend the light of human consciousness deeper into the cosmos. There were moments when the emotion and joy of what he was helping SpaceX accomplish quite literally overcame Rose. He would have to lie down on his bed to process it all.

But everyone has a breaking point. Rose's came after a particularly difficult and emotionally draining meeting with Musk in 2014 over whether Crew Dragon should have touchscreens to control the vehicle's flight (more about this in Chapter 11). He had put in more than five years at SpaceX, but it had felt like twenty. Although Rose was only in his mid-thirties, with a young family, he felt like he'd already had a full career.

"Years on end of 100-hour weeks is really hard on a marriage," Rose said. "If I did this another year, I was going to get divorce papers from my wife. She said the boys needed a father. She needed a husband. She told me to look at my priorities. And she was right. I didn't want my boys to grow up without a father."

Josh Jung arrived at SpaceX in early 2004 as the first full-time engineer in McGregor. He loved the work. Just out of college, he did not have a family, and in building test stands and firing rocket engines, Jung was doing exactly what he wanted to do. The insane hours, often twelve hours a day, seven days a week at a minimum, never stopped during the early years. There might be only one or two test technicians, so when rockets needed testing, no one could take off. They just worked.

Over time, Jung said, the situation has improved. The site went from three people when he started to several hundred employees. So there might be thirty test technicians and thirty ground support technicians. This allowed managers to stagger shifts, with time off. During crunch time, these days, there might still be a month here or there of working twelve hours, seven days a week. This happens when Musk and Shotwell give a manager a project deadline and it's too late to hire more people.

After a decade of helping to lead test stand work in McGregor, Jung was asked to be site director. He didn't want the administrative work, but he took the job to keep the machine going in Texas. After a couple of years, in 2016, he decided to take three months off for what SpaceX calls a sabbatical. During that time, Jung found out he and his girlfriend were pregnant for the first time. He knew he would have to decide: SpaceX or his family.

"My problem is I can't get it out of my head," he said. "When I'm at home, I'm still on my phone, or I'm on my SpaceX computer. I couldn't stop. I liked it too much. My concern was how would I have a family and still do SpaceX? Some people could turn it off, but I couldn't. It's just what I loved. I love the smoke and fire of rockets. We were solving really hard problems. We were doing amazing things. So I thought it would be better for myself if I didn't go back to SpaceX until I figured some things out. Nearly seven years later, I still think about it. There are times when I'm really close to going back."

He hasn't—yet.

After helping to reconstruct Launch Complex 39A for the Falcon Heavy and Crew Dragon launches, Phillip Rench transferred to South Texas to help with Starship in August 2018. He might as well have moved to another planet.

Cape Canaveral and nearby Kennedy Space Center had more than half a century of launching rockets, and over time the combined facility became the largest and most advanced spaceport in the world. Just beyond its gates are small towns and cities with abundant restaurants, hotels, and amenities. Walt Disney World is less than an hour away.

By contrast, the land SpaceX acquired for its Starship launch site in South Texas is probably the most remote beachfront property in the United States on the Atlantic Ocean or Gulf of Mexico. It can only be reached by a two-lane, forty-mile road from Brownsville to Boca Chica Beach. This ragged stretch of pavement, especially when Rench arrived on the scene, feels a little bit like one is driving to the edge of the world. Dry and desolate scrublands accentuate the remoteness of Boca Chica, which stands only a few feet above the Rio Grande River to its south and the Gulf of Mexico to the east. For most of the year, it is blazingly hot. For a few weeks during the winter, it is windy and cold.

SpaceX had only built a few large satellite dishes to help track the Dragon spacecraft by 2018, and a handful of employees, led by John Muratore, worked there. Among Rench's first tasks was moving a mountain of dirt 200 by 300 feet across, standing 26 feet high. SpaceX had arranged for delivery of the dirt years earlier to help prepare the launch site. The pile weighed about 60,000 tons, and this mass slowly sunk into what was essentially a giant swamp. This compressed the surface below to form a nice foundation. Now the dirt needed to be moved to build an actual launch site. Muratore reached out to earth-moving companies and received estimates of more than $1 million to haul away the dirt. But Musk wasn't about to pay that kind of money. So Rench and a few other engineers rented backhoes, bulldozers, and dump trucks, and did the job themselves. It meant weeks on end of insanely repetitive work.

"That basically started everyone on the internet asking, 'What the hell are they doing?'" Rench said.

In early 2019 the South Texas team began fabricating a miniature Starship prototype nicknamed Starhopper. After this stubby vehicle made short test flights in July and August 2019, SpaceX retired it. A few months later, Muratore took a break from SpaceX, and Rench found himself in the role of site director, reporting directly to Musk. It completed a rapid rise for an electronics engineer who had been fixing amusement park rides at SeaWorld in Florida a few years earlier.

Musk would often fly to South Texas on weekends, spending Saturday night at the launch site to check on its progress and prodding where needed. On one Sunday morning, at 2 or 3 in the morning, Rench and Musk were discussing plans to build a Starship factory. Musk liked what he heard so much that he was ready to start. Right then.

"He told me to call the concrete guy," Rench said. "He was like, 'I'm here, so he needs to be here.' And so I called the concrete guy, the company that we used, and of course there was no answer."

This is how Musk gets things done. When he makes a decision, it is implemented immediately. Rightly or wrongly, there are no committees. He is a committee of one. This is a key reason why SpaceX moves so quickly.

"That same weekend he asked me what we really needed to build the Starship factory," Rench said. "So I basically started pointing at all the land around Boca Chica, and I was like I need this, and this, and this, and so on. And he's like, done, done, done, and done. He essentially gave me authorization to build all of the site at that moment."

The pace proved too much for Rench shortly thereafter. For the sake of expediency Musk said the Boca Chica factory should be built beneath three giant tents, each about the size of a football field, nearly 400 feet long and 110 feet wide. Rench had a status call with Musk on the morning of November 22, 2019. It had been a long night for Musk. The prior evening he had hosted Tesla's Cybertruck unveiling event in Los Angeles, which had not gone perfectly. During the presentation, Musk called Tesla Chief Designer Franz von Holzhausen onstage to demonstrate that the vehicle's windows were bulletproof. However, when von Holzhausen tossed a metal ball into the window, it shattered. "Well," Musk said, "maybe that was a little too hard." So the next morning, Musk was not in the best of moods. Especially when Rench reported that the lead time on obtaining and installing the tents was longer than expected.

"I don't think he had slept," Rench said. "And he threatened to fire me, over the phone, for something really silly. Over the tents. I was just so burnt out at that point. I was working eighty or ninety hours a week, and I hadn't seen my family in a month, if not longer. I had hit my full burnout point."

After the phone call with Musk, Rench telephoned Zach Dunn. He was done and wanted to get about as far away from rockets as he could. He and his wife, Gwendolyn, moved to southern Maine and bought a twenty-acre farm where they planted about 100,000 fruit trees. In the summer, they invite the local community to come and pick strawberries. He happily traded the red planet for red fruit.

Over and over again, employees who rose to the upper echelons at SpaceX said they had to accept that they were living on borrowed time.

"If you are a director at SpaceX, and certainly if you are a VP, you need to mentally accept that you are already dead," said Abhi Tripathi, who worked at the director level for five years. "This sounds dramatic, but every VP has a near 100 percent chance of being fired or completely burning out. This is a double-edged sword in that you are under a lot of pressure, but you are also liberated."

Only Gwynne Shotwell, so essential to SpaceX customers and talented at working alongside Musk, has defied this gravity. Hans Koenigsmann, who worked closely with Musk for nineteen years at SpaceX, longer than anyone but Shotwell, said the key was striking the right balance between pacing oneself and managing Musk's expectations. The best of these managers also provided top cover for their teams, and Koenigsmann is almost universally respected by everyone who worked with him because he labored to ensure his employees were not sucked into the machine. As for directors and VPs, he agreed that to succeed one had to be willing to accept that firing was inevitable.

"The way to success as a manager was to not rely on your job, in my opinion," Koenigsmann said. "That makes you much more valuable as a manager for the company, because you aren't trying to save yourself constantly."

What are we to make of the SpaceX work culture?

The cost of Mars is high. For employees, it is a choice as to whether they're willing to pay it. Taking a job at SpaceX means working "hard core" on some of the most challenging engineering problems in the world—and beyond. After Musk bought the social networking site Twitter in October 2022 for $44 billion, he similarly asked the staff there to work hard core on cutting costs and improving the product. This meant long hours and mass layoffs for staff Musk deemed nonessential. The results of Musk's management of Twitter—renamed X—might politely be described as uneven.

The reason why Musk's style works at SpaceX, but not X, is simple. The 10,000 people who chose to work at SpaceX knew what they were getting into. Musk is a known quantity in the space industry. Prospective employees speak with friends about the work environment. Most importantly, they believe in the mission. Like, they *really* believe. And their vision aligns with Musk's sweeping and passionate goals for spaceflight. Robert Rose, after all, said he would step on a landmine for Musk. Not everyone felt so fervently, but most SpaceX employees were true believers. By contrast, many Twitter employees had never signed up to work for Musk and were appalled by their new boss's vision for their social network and his determination to rapidly bring it about.

Lori Garver has known Musk for two decades. She was an ally of his during the Obama White House, when she served as NASA's deputy administrator. But she also cares deeply about a healthy aerospace workforce, especially for women and minorities.

"If I thought some part of working at SpaceX was inhumane, I would have elevated that concern," Garver said. "It's not sustainable for your whole life. They do wear through people. The environment at SpaceX is, if you want to be here, you're going to work really hard. And everybody knows that going in. That doesn't mean there isn't room for improvement."

There is also room for improvement when it comes to diversity. Alarmed at the lack of women in the space industry, Garver teamed up with Will Pomerantz and Cassie Lee to found the Brooke Owens Fellowship in 2017. They'd all known and loved Dawn Brooke Owens, a pilot and space policy expert, who died of breast cancer at the age of thirty-five. Every year, from about one thousand applicants, the organization chooses fifty undergraduate women and gender-minority students for paid internships and executive mentoring at space companies. Three years later, Pomerantz cofounded the Patti Grace Smith Fellowship for Black undergraduates seeking an aerospace career.

Even today, despite all of the negative headlines about Musk and the work environments at his businesses, SpaceX continues to be the company most requested by applicants to those fellowships. Garver said she talks with "Brookies" all the time about their experiences. Some companies have been removed as hosts because of problems. But the interns placed at SpaceX gush about their experience.

"Many stay and work there," she said. "They are thrilled for their jobs. They work hard because they love it. They have given up social lives because they care about it. I'm not sure how long that is sustainable, but they aren't being abused."

SpaceX has had its fair share of labor disputes, harassment concerns, and other issues. Some of these are simply endemic to large companies, but others are related to Musk's behavior. The most damning of these incidents occurred in 2016 aboard one of Musk's private jets, involving a flight attendant working on a contract basis. The details of the alleged indecent exposure have not been publicly confirmed, and according to *Business Insider* Musk, SpaceX, and the flight attendant entered into a severance agreement in 2018 granting the attendant a $250,000 payment in exchange for a promise not to sue over the claims. The deal included a nondisclosure agreement. Publicly, Musk said the incident "never happened."

After this became public in 2022 it understandably stirred discontent at SpaceX. Seeking to calm the waters, Shotwell wrote in a company-wide email that "I have worked closely with him for 20 years and never seen nor heard anything resembling these allegations." What happened next would prove revealing about the company's corporate culture and the cost of Mars. Several employees wrote an open letter to senior management raising concerns about Musk's behavior and its negative impacts on the company. "SpaceX must swiftly and explicitly separate itself from Elon's personal brand," the letter stated. Hundreds of employees signed it, mostly anonymously. Instead of fostering change, however, the letter antagonized senior leaders at SpaceX. Shotwell responded that employees should focus on their jobs. Ultimately nine people were fired, at least in part for their involvement with the open letter.

The message was crystal clear: to work for SpaceX meant to work for Elon Musk. The company and the man, for better or worse, were inseparable.

## "This thing better look like a goddamn beehive 24/7."

In the weeks following the dramatic AMOS-6 explosion in 2016, I distinctly remember thinking that Musk misread the room when he pressed ahead with his speech about Mars. At the time America's relationship with Russia had already turned frosty following the takeover of Crimea. Russia kept raising the price of the Soyuz seats that NASA astronauts flew in to reach the International Space Station, all the way to $81 million. Through its Commercial Crew program, NASA was funneling billions of dollars to SpaceX and Boeing to break its dependence on Russia for astronaut flights.

I spoke with a lot of NASA people at the time, and the sentiment boiled down to this: Musk's dream of sending thousands of people to settle Mars may just have to be put on the back burner until he could send two people to the space station. I wrote an article stating as much. Musk, of course, did not listen to me or anyone else who suggested he might focus a little more on getting the Falcon 9 flying safely, and a little less on far-flung dreams and super-rockets.

By and large, NASA and SpaceX have had a fantastically fruitful relationship for nearly two decades. The U.S. space agency funded Musk's company at critical moments, and in turn SpaceX has provided the lowest cost, fastest, and often the best solutions to NASA's spaceflight needs. But there has often been an underlying tension. In seeking to send humans to Mars, SpaceX is outflanking NASA's primary mission of exploring the solar system and beyond. These aggravations increased when Musk put these far-out goals ahead of his most important customer's immediate needs—getting astronauts to the space station on Crew Dragon.

Three years after the Guadalajara speech, these tensions boiled over again. It was September 2019, and SpaceX employees were working feverishly to complete the very first full-size prototype of Starship, named Mk1. They faced a deadline of September 28, when Musk had scheduled a showy public unveiling in South Texas. Teams in California pushed through a schedule with tasks, marked out by the hour, that they needed to complete for the event to go off as planned.

In the final weeks before the event, Musk convened daily meetings in his executive conference room in Hawthorne, with about two dozen engineers seated and standing. On the phone, he had leadership of the launch site team in Boca Chica. At the outset of one of these meetings, Musk gave his senior leaders a clear directive.

"You have one objective," he said. "Get this thing done as fast as possible."

To make sure of this, Musk said he wanted a camera to be set up at the build site, allowing him to observe progress on the 160-foot-tall vehicle.

"This thing better look like a goddamn beehive 24/7," Musk said. "If it doesn't look like a goddamn beehive 24/7, that's fucked. That's why I'm asking for a time-lapse camera. I really just want to know, does this look like a busy beehive? Because if it doesn't look like a busy beehive twenty-four hours a day, that's fucked up. If we need to hire more people, or get more contractors, we should get that. And if you say, well, I don't know how to do it, let me know. I'll help you. But don't not do it."

Later, during a meeting on September 13, Musk offered to directly call the chief executives of welding companies that SpaceX had contracted with. It was late on a Friday night, so Musk agreed to call the next morning. But he said he would ask the CEOs to set aside all of their other projects to focus on Starship. He was willing to pay the extra costs of pulling welders off existing work.

"We literally want their best people to stop working on whatever they're working on, and come work on our stuff," Musk said. "Don't send the B-team, send the A-team. Send them to Boca Chica. Probably some of their customers might be a little mad, but I think they will not be super angry if they know this is for humanity's future in space. We appreciate that your brewery might have its fucking thing two weeks late but consider one versus the other."

Musk was focusing the totality of SpaceX's effort on completing the Starship prototype for an event planned on a starry Saturday night in South Texas later that month. He was willing to spend lavishly to scrounge up welders and builders and whoever else was needed to finish a vehicle that would never launch, for an arbitrary deadline and a glitzy showcase.

This frantic focus on Starship rankled the NASA administrator at the time, a former fighter pilot and congressman named Jim Bridenstine. SpaceX and Boeing were two years behind their original schedules, and each still faced serious challenges before flying crew. NASA kept having to go to the Russians, hat in hand, to buy seats for its astronauts. And here was Elon bloody Musk moving heaven and Earth for Mars, once again, while ignoring NASA.

Bridenstine couldn't help but give vent to his emotions. On the day before Musk's ballyhooed Starship unveiling, Bridenstine tweeted some words that reflected his frustration with what he viewed as SpaceX's misplaced priorities.

"I am looking forward to the SpaceX announcement tomorrow," Bridenstine said. "In the meantime, Commercial Crew is years behind schedule. NASA expects to see the same level of enthusiasm focused on the investments of the American taxpayer. It's time to deliver."

SpaceX would.

| 11 |

# THE FABERGÉ EGG

*December 2008*
Washington, D.C.

The Commercial Crew program was a decade old when Jim Bridenstine tweeted "It's time to deliver." At its beginning, almost no one in Washington, D.C., much liked the idea of private companies flying NASA astronauts to the International Space Station. Only government space agencies were thought capable of such a sacred task. Even the architect of the Commercial Cargo program, NASA Administrator Mike Griffin, viewed crew flights as a bridge too far. Once private companies like SpaceX had "amply demonstrated" cargo delivery, he said, they might be considered for human spaceflight. But not before.

Then the Great Recession of 2008 cracked the door open. Barack Obama had been elected president, and the outgoing Bush administration deferred the crafting of an economic stimulus to the incoming president. An ally of commercial space, Lori Garver led Obama's transition team for NASA, and the new administration was attentive to her pitch about the potential for commercial space to lower spaceflight costs and revitalize the U.S. launch industry. But Garver knew Congress would not easily accept the idea with influential voices like Griffin in opposition and with traditional space companies seeking to block new entrants such as SpaceX from receiving a greater slice of NASA funding. So, to start the Commercial Crew Program she asked the Obama White House to tuck $50 million into the must-pass $800 billion American Recovery and Reinvestment Act.

"We knew the recovery act was going forward, and Congress would not have the opportunity to nitpick it," Garver said. "Maybe it was a little bit sneaky, but otherwise it probably was not going to happen."

Griffin and his allies in Congress knew the space shuttle, America's only means of putting humans into orbit, would retire soon. But they believed NASA could rely on Russia's Soyuz vehicle in the near term, and the space agency's plan to develop a rocket to fly the Orion spacecraft to low-Earth orbit in the long term. In hindsight, both were terrible ideas. The Soyuz was reliable for a time, but by the end of the 2010s Russia's space program began to have serious quality control issues. And Russia's invasion of Ukraine would have put NASA in a terrible position had the Soyuz been America's only lifeline to the station. As for Orion, it will not fly humans into space before at least 2025, with a per-seat cost ten times that of Commercial Crew.

After the recovery act's passage in February 2009, the new program got a big break. Boeing, the most trusted name in U.S. spaceflight, agreed to bid for Commercial Crew contracts. This eliminated one argument of naysayers, who said private companies were not up to the task. Critics could not credibly argue that Boeing, NASA's most important contractor, would fail at human spaceflight. Boeing's entry started to break the dam of congressional funding, and by the fall of 2010 a few hundred million dollars started flowing.

## How should a modern spaceship land?

Over the next four years the competition boiled down to three players: Boeing, SpaceX, and a Colorado-based company building a spaceplane, Sierra Nevada Corporation. Each had its own advantages. Boeing was the blueblood, with decades of spaceflight experience. SpaceX had already built a capsule, Dragon. And some NASA insiders nostalgically loved Sierra Nevada's Dream Chaser space plane, which mimicked the shuttle's winged design.

This competition neared a climax in 2014 as NASA prepared to winnow the field to one company, or at most two, to move from the design phase into actual development. In May of that year Musk revealed his Crew Dragon spacecraft to the world with a characteristically showy event at the company's headquarters in Hawthorne. As lights flashed and a smoke machine vented, Musk quite literally raised a curtain on a black-and-white capsule. He was most proud to reveal how Dragon would land. Never before had a spacecraft come back from orbit under anything but parachutes or gliding on wings. Not so with the new Dragon. It had powerful thrusters, called SuperDracos, that would allow it to land under its own power.

"You'll be able to land anywhere on Earth with the accuracy of a helicopter," Musk bragged. "Which is something that a modern spaceship should be able to do."

A few weeks later I had an interview with John Elbon, a long-time engineer at Boeing who managed the company's commercial program. As we talked, he tut-tutted SpaceX's performance to date, noting its handful of Falcon 9 launches a year and inability to fly at a higher cadence. As for Musk's little Dragon event, Elbon was dismissive.

"We go for substance," Elbon told me. "Not pizzazz."

Elbon's confidence was justified. That spring the companies were finalizing bids to develop a spacecraft and fly six operational missions to the space station. These contracts were worth billions of dollars. Each company told NASA how much it needed for the job, and if selected, would receive a fixed price award for that amount. Boeing, SpaceX, and Sierra Nevada wanted as much money as they could get, of course. But each had an incentive to keep their bids low, as NASA had a finite budget for the program. Boeing had a solution, telling NASA it needed the entire Commercial Crew budget to succeed. Because a lot of decision makers believed that only Boeing could safely fly astronauts, the company's gambit very nearly worked.

The three competitors submitted initial bids to NASA in late January 2014, and after about six months of evaluations and discussions with the "source evaluation board," submitted their final bids in July. During this initial round of judging, subject-matter experts scored the proposals and gathered to make their ratings. Sierra Nevada was eliminated because their overall scores were lower, and the proposed cost not low enough to justify remaining in the competition. This left Boeing and SpaceX, with likely only one winner.

"We really did not have the budget for two companies at the time," said Phil McAlister, the NASA official at the agency's headquarters in Washington overseeing the Commercial Crew program. "No one thought we were going to award two. I would always say, 'One or more,' and people would roll their eyes at me."

The members of the evaluation board scored the companies based on three factors. Price was the most important consideration, given NASA's limited budget. This was followed by "mission suitability," and finally, "past performance." These latter two factors, combined, were about equally weighted to price.

SpaceX dominated Boeing on price. Boeing asked for $4.2 billion, 60 percent more than SpaceX's bid of $2.6 billion. The second category, mission suitability, assessed whether a company could meet NASA's requirements and actually safely fly crew to and from the station. For this category, Boeing received an "excellent" rating, above SpaceX's "very good." The third factor, past performance, evaluated a company's recent work. Boeing received a rating of "very high," whereas SpaceX received a rating of "high."

While this makes it appear as though the bids were relatively even, McAlister said the score differences in mission suitability and past performance were, in fact, modest. It was a bit like grades in school. SpaceX scored something like an 88, and got a B; whereas Boeing got a 91 and scored an A. Because of the significant difference in price, McAlister said, the source evaluation board assumed SpaceX would win the competition. He was thrilled, because he figured this meant that NASA would have to pick two companies, SpaceX based on price, and Boeing due to its slightly higher technical score. He wanted competition to spur both of the companies on.

The decision was to be made on August 6, during a meeting at NASA headquarters. The agency's head of human spaceflight, Bill Gerstenmaier, convened his top human spaceflight advisors in the agency's "Space Operations Center" at headquarters. This secure room was built after the *Columbia* accident in 2003 for high-level strategic meetings. Gerstenmaier and about twenty senior officials at NASA sat around a large, rounded table, discussing the source evaluation board scores with the aim of picking a winner.

After a presentation on the technical scores, Gerstenmaier asked each advisor for an opinion. These were the who's who of the U.S. spaceflight community, many of whom, like Gerstenmaier, had come up in the Space Shuttle Program, long before the era of commercial space. As he went around the room, each person echoed the same response, "Boeing." First five people, then ten, and then fifteen. This seemed to please Gerstenmaier, known warmly as "Gerst" in the global spaceflight community, and encouraged potentially dissenting voices to fall in line. McAlister watched this cascade of pro-Boeing opinions sweep around the table, a building and unbreakable wave of consensus, with mounting horror.

"I'm freaking out because I could see them going with Boeing, which in my opinion was an inferior proposal, and only with Boeing," he said. "It was not groupthink; it's just that everyone at the time was comfortable with Boeing. SpaceX had only been flying cargo to the space station for two years."

Having only come to NASA in 2005, McAlister was not part of this human spaceflight firmament. Sitting to Gerstenmaier's right and reporting to him as the director of commercial space development, McAlister's mind whirled with possibilities for throwing himself in front of the oncoming bus. He knew that arguing SpaceX had presented the best proposal, based on price and technical merit, would get him nowhere.

Near the end of the discussion, Gerstenmaier solicited McAlister's opinion. In turn, McAlister started asking questions. First, he turned to Bill McNally, the agency's head of procurement. Prior to joining NASA, McNally had spent nearly three decades in acquisition for the U.S. Air Force, managing the Tomahawk cruise missile program and leading technology contracts for the "Star Wars" missile defense program. McAlister asked if the veteran procurement official had ever seen a federal agency choose a bidder that cost 60 percent more when both bids were technically acceptable.

McNally shifted uncomfortably in his seat at this question. Eventually he remarked that the source selection official, Gerstenmaier, could do whatever he pleased. McAlister pressed further, repeating the question. "No," McNally replied. "That would be uncharted territory."

Next, McAlister questioned the engineer representing safety and mission assurance, Deirdre Healey. When she had spoken, Healey said the safety division preferred Boeing as long as the company performed an in-flight test of its spacecraft's abort system—powerful thrusters that push the vehicle away in case the rocket malfunctions during launch. But Boeing did not plan to do so. Their bid included a ground test of this abort system, not one in flight. McAlister seized on this, asking Healey if this meant Boeing's proposal should really be considered unsatisfactory.

No, Healey replied, indicating the bid was acceptable to her.

Another member of the source evaluation board at the meeting, a deputy procurement manager from Johnson Space Center named Lee Pagel, said this question scored points for McAlister. It was strange that so many smart people thought NASA could just snap its fingers and Boeing would conduct an in-flight abort test. "In all my years of working with Boeing I never saw them sign up for additional work for free," he said.

After addressing his questions to McNally and Healey, McAlister turned to Gerstenmaier.

"I told Gerst he had to pick two," McAlister said. "His head of safety and mission assurance just said Boeing's proposal was unsatisfactory, and the head of procurement said the cost would be difficult to defend. And Elon sues everybody."

Typically, a decision is made at this meeting. But Gerstenmaier said he needed to think about all he had heard. He took another month. During this timeframe someone at NASA floated the idea of a "leader" and a "follower," with Boeing getting the lion's share of funding and SpaceX a small amount to keep going. But Musk rejected this immediately.

At the same time, McAlister kept pushing Gerstenmaier, telling him competition was essential to moving the program forward as Boeing and SpaceX strove against one another to build the safest, most reliable, and most cost-effective system. Eventually, Gerstenmaier agreed. He called the NASA administrator, Charlie Bolden, to say he was going to blow a hole in the agency's budget. Instead of asking Congress for $870 million in the budget for Commercial Crew the next fiscal year, NASA would need $1.25 billion.

It had been a very near thing. NASA officials had already written a justification for selecting Boeing, solely, for the Commercial Crew contract. It was ready to go and had to be hastily rewritten to include SpaceX. This delayed the announcement to September 16.

Seven years later, when Russia invaded Ukraine, tensions between NASA and Roscosmos, the Russian space agency, exploded. The pugnacious leader of Russia's space program, Dmitry Rogozin, blustered that he would kick NASA off the station. But because the U.S. space agency had opted for competition between SpaceX and Boeing in its crew program, this was a hollow threat. Dmitry would dance with Dragons, and get burned.

## NASA hated it

The Dragon team at SpaceX felt uneasy about the outcome of the crew contract before its announcement. David Giger helped write the proposal, working closely with former NASA astronaut Garrett Reisman, who joined SpaceX in 2011 to help land contracts for Dragon and provide an astronaut's perspective. But their small team was no match for

Boeing's proposal-writing machine. It was intimidating knowing that 200 people were working on Boeing's proposal, when Dragon's team could fit in a small conference room.

All along NASA had said it wanted to maintain competition, but in the summer of 2014 Giger and his colleagues started to hear that NASA would probably only pick one, and how NASA preferred a winged vehicle that could land on a runway. They feared that Boeing would be the first choice, and if NASA picked a second winner it would take a flier on Dream Chaser. The Dragon team's stress multiplied when a decision they expected on the morning of September 16 did not come, dragging out into the afternoon hours.

"We were on pins and needles," Giger said. "Finally, Gwynne got the message from NASA, and she told us. And it was like holy shit, now we have to do this."

One of the big challenges involved the flight abort system. Unlike Boeing, SpaceX told NASA it would perform both a ground-based test of Dragon's SuperDraco thrusters and one in flight, with Dragon powering away from a rocket. Crew Dragon had eight of these incredibly powerful thrusters. Each provided 16,000 pounds of thrust, or nearly 200 times the thrust of Dragon's smaller Draco thrusters. The first test, called a "pad abort," was scheduled for May 2015. At Cape Canaveral, a prototype spacecraft nicknamed *DragonFly* would fire its SuperDracos for seven seconds, reaching an altitude of about a mile and a half, before splashing down offshore in the Atlantic Ocean.

Just days before the test, the propulsion team, led by an engineer named Matt McKeown, discovered that valves controlling the flow of helium into the propellant tanks—this helium pressurized propellant before it moved into the engine—were sometimes closing prematurely. As a precaution, on the night before the test, McKeown told Giger he wanted to send a repeat command to the valves one second after Dragon's propulsive burn started. This would reopen any shut valves. The Dragon team spent the night rewriting the flight software and testing it.

The next morning, on May 6, *DragonFly* blasted off the pad. As McKeown feared, some of the tank pressurization valves shut prematurely. A few seconds into the flight, a subset of the prototype Dragon's thrusters shut down, producing a cloud of orange smoke from excess oxidizer, rather than a smooth burn. But the extra valve command saved the day. Had SpaceX not changed the software, Dragon would have petered out after a second or two of thrust. In this case, the command to reopen the valves kept Dragon flying long enough to reach about half of its target altitude, 3,800 feet.

This was high enough to get the data needed for the test. But still, Giger worried Dragon's momentum would not carry it far enough offshore. The launch pad lay only a few hundred yards from the coast, and the nearshore waters were shallow. A hard landing might destroy Dragon and produce a huge fireball. But in the end, Dragon made it just far enough. Had it landed a mere fifty feet closer to shore, the spacecraft would probably have been lost. As it was, Dragon had to be dragged into deeper water before the recovery ship could pick it up.

After plucking *DragonFly* from the Atlantic, SpaceX shipped it to Texas for a series of flight tests to validate the ability of the spacecraft to propulsively land. For Musk, this was the centerpiece of Dragon's development. Fundamentally, it was a capsule like NASA and the rest of the world had been building since the 1960s. By propulsively landing Dragon, however, SpaceX could usher spaceships into the twenty-first century. Moreover, Musk pushed it hard because he knew the Martian atmosphere was too thin for large spacecraft to land with parachutes. Eventually, the company would need powered landings, and he wanted to start with Dragon. With much fanfare, SpaceX even announced a

"Red Dragon" plan to send an uncrewed Dragon spacecraft to Mars by as early as 2016.

While this raised exciting possibilities for science on Mars, it chilled the hearts of NASA officials charged with safely getting astronauts to and from orbit. Propulsive landing was not explicitly part of SpaceX's Commercial Crew bid to NASA in 2014, but Musk intended to eventually alter the agreement to allow it in lieu of parachute landings at sea. However, no one at the space agency wanted propulsive landing, least of all the astronauts who would ride inside. "NASA hated it," McAlister said. "We knew it was going to be very, very hard and lead to delays."

Propulsive landing did prove to be very, very hard. Typically, with a capsule-based design, there are two elements. There is a crew compartment, or top of the vehicle, where the astronauts live. And then, at the aft end, there is a service module, which houses solar panels and propulsion, including propellant tanks and an engine. The service module powers the spacecraft throughout its journey, and then is jettisoned before reentry into Earth's atmosphere. After separating, it burns up while the crew capsule flies home safely. One problem with this design approach is that SpaceX did not want to throw away a service module with a valuable propulsion system.

SpaceX, therefore, built the Draco and SuperDraco thrusters into the crew capsule, sitting atop a mostly empty "trunk" that could haul cargo to the space station but would not be overly expensive to replace. This added complexity to the crew vehicle, however.

Draco thrusters sipped fuel as if through a straw. But SuperDraco thrusters guzzled fuel that issued like the torrent from a fire hydrant, ingesting a flow more than 200 times faster. In a traditional spacecraft, with a service module, these functions might have been split into separate fuel tanks. But Dragon crammed all of this into the structure of its crew compartment. And because the thrusters shared common tanks, special propellant management devices had to be built to gather fuel in zero gravity for the thrusters, while also operating under the high g-forces of atmospheric reentry.

That was only the beginning of the challenges. Rocket engines need to "throttle," or change their overall thrust up or down, like an automobile engine controlled by a gas pedal. After much trial and error with the Merlin rocket engine, SpaceX was able to increase the dynamic range of the Merlin engine from 100 percent power down to 60 percent. With the SuperDracos, the range had to be much greater, with 100 percent power needed for an abort and just 10 percent for landing. The solution, Giger said, was a "crazy throttling valve." Dragon also needed extendable legs and a host of other new technology to accommodate landing.

As Giger's team engineered solutions to these problems, it added layers of complexity to the Dragon spacecraft. The propulsion system and its intricate propellant tanks, in particular, were the subject of many heated meetings with Musk. His hard-charging, go-faster approach got bogged down in a thicket of technical challenges. Dragon, so fragile and exquisite in its design, was different from rockets. It could not be brute-forced. Over time, Musk and the engineering team began to refer to Crew Dragon as the Fabergé egg, after the opulent Easter gifts crafted for Russian tsars. Eventually, with unlimited time and money, Giger's team could have made it work. But they had neither the time nor the money.

Musk's determination to press ahead with propulsive landing started to falter in 2016 after several discussions with Kathy Lueders, who had gone on to lead NASA's Commercial Crew program. Her philosophy toward SpaceX and Boeing was to be as hands-off as possible. But as she watched SpaceX lose more than a year of development time chasing powered landing, she asked her team to gauge how much testing and certification it would take for the government to accept it as safe to land astronauts with SuperDracos.

"I've always respected Elon's need to make the technical decisions," she said. "When it's your spacecraft, it's your decision. But I was able to talk to him and quantify how much work it would take to qualify this. I wasn't going to tell him what to do, but we talked about the time frame."

Lueders knew Musk wanted to beat Boeing and its Starliner spacecraft to the space station with astronauts. SpaceX had so much other work to do before Dragon could fly crew, from life support systems to spacesuits. Lueders explained that by pursuing propulsive landing, SpaceX probably would lose the race to Boeing because it would be distracted from all of this other work.

The final straw came when the Dragon team determined there was no way to avoid adding parachutes to the vehicle. While SuperDracos could be the primary means of landing, the entire propulsion system could not be made to survive two failures while still functioning, known in aerospace parlance as "two-fault tolerant." NASA would not give SpaceX an exception to this rule for propulsive landing, either. So Dragon was flying with chutes one way or the other. Musk decided in 2017 that Dragon would not, in fact, land under its own power. In an interview a couple of years later, he told me that it had been a painful but necessary decision.

"The reason we did propulsive landing development on Dragon was for Mars," he said. "It was not for Earth. On Mars, you cannot land purely with parachutes. The thought was we could extend Dragon to Mars. That's why we put so much effort into a propulsive landing system for Dragon. This caused a massive amount of grief for the Dragon development team, and in retrospect, I feel guilty about that."

The challenge of propulsive landing became the Starship program's problem. All the painful lessons learned in Dragon would feed into making a smarter, more capable spacecraft. As part of that, Musk took the Fabergé egg experience to heart and brought a new mantra into the development of Starship. Whereas Dragon's propulsion tanks had been complex, Starship would aspire to simplicity in its design.

Over and over again in meetings Musk would say, "The best part is no part."

## Touchscreens, space tuxedos, and other Crew Dragon concerns

Powered landings were not the only sticking point in SpaceX's contract with NASA for Crew Dragon. Another pressing issue involved the controls by which astronauts would fly the spacecraft. Virtually every spacecraft built before Dragon had translational and rotational hand controllers, or joysticks, for manual control of the vehicle.

But within the company a heated debate raged about how Dragon should be flown. Robert Rose, the director of flight software, felt strongly that astronauts should not have direct control of the vehicle. It would introduce too much complexity and add months, if not years, to the testing and development of Dragon. He advocated for a touch-screen interface similar to a tablet, as well as restrictive flight plans. SpaceX's in-house astronaut, Reisman, pushed just as forcefully in favor of manual controls. In meetings, he would passionately argue that any-one brave enough to sit on the pointy end of a rocket should have con-trol of their fate.

Reisman eventually came around to flying without manual controls, but he opposed Rose's idea of predetermined waypoints on the touch-screen interface. That is, astronauts could not turn the vehicle right or left on command, but had to select from a few paths that were offered by the flight software. This became a recurring battle between Houston and Hawthorne, with NASA flight operators arguing for more extensive backup capabilities and SpaceX preferring automation. The skirmishes extended to critical functions, such as giving astronauts the ability to manually deploy the parachutes, or ignite SuperDracos during ascent to de-orbit Dragon immediately.

"Elon and most of SpaceX wanted to make a fully automated vehicle and were afraid to give the crew control over any of these functions for fear that they would make an error," Reisman said. "The NASA operations community wanted much more extensive manual backup capabilities. I was stuck in the middle. Working out a mutually agreeable compromise was my most difficult task during my time at SpaceX."

At Musk's direction, as part of SpaceX's bid in 2014, Crew Dragon was slated to include manual controls. But this was not the final word. After SpaceX won the crew contract, its engineers put together three options for flying Dragon. One involved traditional translational and rotational hand controllers, the second choice was a Nintendo- or PlayStation-like controller, and the third a touchscreen. A handful of astronauts flew out to Hawthorne to test these options, and the conclusion was that, while the astronauts preferred a joystick, the touchscreen proved to be functional as well.

This helped lead to a compromise. From an aesthetic standpoint, Musk felt screens were a better fit for a modern spacecraft. He sided with Rose for this type of interface. But he agreed with Reisman that the waypoint system was untenable, and that astronauts should be able to control the flight of Dragon in space. Reisman then had to sell this compromise to NASA.

Musk's initial decision to opt for joysticks in the Commercial Crew proposal is illustrative of how he often deals with NASA and other government agencies. It was a bit of realpolitik to mollify fight controllers and the influential astronaut community. Had SpaceX bid a contract that lacked joystick controllers, it would have given the Astronaut Office at Johnson Space Center an easy and straightforward reason to lobby against the SpaceX proposal.

The joystick battle simmered on the back burner for a few years while the Dragon team worked on more pressing matters, like propulsive landing, parachutes, and other systems. But by 2018 the issue returned when SpaceX invited several astronauts to evaluate a new, upgraded touchscreen interface to fly Dragon and dock with the space station (in a simulation, of course). Half a dozen astronauts tried, and they all struggled. Only one of the six succeeded on the first attempt, a shuttle pilot named Doug Hurley.

It's not that Hurley found the spacecraft easy to fly in the simulation. Far from it. But he was a former fighter pilot and a veteran of two space shuttle missions, where he had flown inside a rendezvous simulator hundreds of times. Hurley only managed to dock Dragon because of his long experience as a test pilot and with the shuttle simulator experience.

"Frankly, Dragon flew like shit," he said. "It really was not flyable at all. If you've got six astronauts, and five of them can't do a task, then you've got a problem." So the Dragon team went back to work.

The astronauts also needed something to wear. When a spacecraft launches and returns to Earth, humans wear a pressurized suit for a couple of reasons. One involves the loss of cabin pressure, in which an air leak in the spacecraft might not destroy the ship. In this case, a pressure suit would give crew members a fighting chance. Suits also provide backup safety mechanisms.

Musk, however, found every spacesuit he had seen before to be clunky and unattractive. According to Giger, he wanted the SpaceX spacesuit to feel like a tuxedo: "Elon must have said this a thousand times. 'When people put a spacesuit on, I want it to look like when people wear a tuxedo. Anyone can wear a tux. You can be tall. You can be short. You can be big. But if you put a good-fitting tux on, you look good. I want that to be the case with the suit.'"

For one thing, Musk disliked the "rubber boots" that NASA and Russian astronauts wore riding the space shuttle and Soyuz vehicles into orbit. He wanted SpaceX's suit to have boots that did not look like plain-old rubber boots. And of course, he did not want SpaceX to spend a fortune on spacesuits, even though every astronaut needed a custom-tailored suit. During the sizing process, about fifty measurements are taken and fed into an algorithm that determines an ideal suit size. Musk said tailored spacesuits should cost about $10,000, the price of a high-end, custom (nonspaceworthy) tuxedo.

The world leader in the design of aviation and spacesuits is the Massachusetts-based David Clark Company, which dating to the beginning of World War II has supplied nearly every U.S. high altitude and space program. The firm was already working with Boeing on its Starliner flight suits when SpaceX asked for a quote. The price came back at nearly $1 million a suit. That was a nonstarter, so Reisman and the suit team, led by Jason Tenenbaum, had to get creative. For the suit's gloves, which are often the most difficult component, Tenenbaum hired a mechanical engineer named Peter Homer who had independently won two NASA challenges for glove design. Homer's gloves, initially designed for the fun of it, were made entirely of fabric for maximum dexterity even during pressurization of the suit. This basic design is flying today.

SpaceX also pushed to additively manufacture the helmets and had to win over skeptics about the quality of 3D-printed material—some NASA engineers were not thrilled by the prospect of plastic protecting the heads of astronauts. But by printing the helmet, the contouring and sizing could be easily customized. In the end, Tenenbaum and his team whittled the price down to a small fraction of what David Clark had asked for. This is fairly typical for SpaceX. The price was substantially higher than what Musk wanted, but also substantially lower than the industry standard.

That left just the style of the suit. SpaceX went through more than fifty designs and nearly as many meetings with Musk. The company hired a number of different designers to get the aesthetics just right.

Finally, at a meeting in 2015, a model walked into the conference room wearing the latest design, a black and white suit that looked decidedly futuristic with its sleek lines and angles.

"Okay," Musk said. "That's it."

## Musk versus the entire human spaceflight community

As the Dragon teams grappled with myriad technical problems, SpaceX also confronted a mountain of skepticism within the human spaceflight community. The stakes were just so much higher with the lives of NASA astronauts on the line, and this opened up SpaceX to a much broader array of scrutiny.

One legacy of three fatal spaceflight accidents in NASA's history (the Apollo 1 fire and two space shuttle losses) is a multitude of independent review boards that dig into the agency's safety protocols. In the summer of 2011 SpaceX was invited by one of these, the Aerospace Safety Advisory Panel, to speak about the Dragon spacecraft's software. Shotwell thought it best to decline, but McAlister and Lueders talked her into participating. They believed that, if the panel heard directly from SpaceX, it might tamp down some of its members' apprehension about the upstart company.

Shotwell relented and asked Rose to provide an overview of SpaceX's flight software. He brought some technical slides with him to the meeting in Houston, but as they waited to address the panel, Shotwell urged

Rose to talk about his past experience coding video games for PlayStation. They'll love that, she said.

Spoiler alert: the panel, chaired by Navy Vice Admiral Joseph Dyer, did not love that.

"What I didn't realize at the time is that we were basically walking into a firing squad," Rose said. "Against my better judgment, I started telling them about my background. I get interrupted, and someone says, 'Robert, this isn't *Space Invaders*.' So I look at Gwynne with this sort of help-me expression, and she gives me this firm nod to keep going and power through it."

From that point on the panel started needling Rose with questions and concerns. It was like a comedy club when the crowd decides an act is not funny and turns on the comedian. But Rose did try to power through, and eventually got into a fairly technical discussion about fault management software. He mentioned a specific test coming up and, in his nervousness, forgot his audience lacked his technical depth. Rose erred when he started to describe details of the test and how it was designed to ensure there were no mistakes in coding the fault detection and fault response software.

As he spoke, one of the panelists interrupted him: "So let me get this straight, you're saying you have no mistakes in your software?"

It was such a ridiculous accusation that Rose stood there dumbfounded for a few seconds trying to figure out how to respond. He tried to clarify what he meant but kept getting interrupted. Half of the panelists started to energetically talk amongst themselves, and the other half lectured Rose. They wagged their tongues and their fingers at him, recounting tales of yore about famous software bugs. As if Rose, the software programmer, had never heard of them. It began to dawn on him that this had been a trap, and he had walked right into it.

Rose's embarrassment was compounded two months later when the safety panel published a public summary of its August meeting. "The SpaceX software presentation was unsettling to the review team," the summary stated. "Their comments with regard to software were very disturbing and presented a lack of insight and sophistication in what can go wrong in this business."

In its first real interaction with SpaceX on these issues, the NASA safety community had come away unimpressed. "We felt that letting the ASAP panel hear directly from SpaceX would be helpful," McAlister said. "Turns out, not so much. It was a disaster."

SpaceX would have another, more serious run-in with a different safety panel, the NASA International Space Station Advisory Committee, four years later. It concerned propellant densification and SpaceX's need to launch the Falcon 9 rocket soon after fueling the vehicle with super-chilled fuel and oxidizer. This meant SpaceX would strap the crew into Dragon and then fuel the rocket with astronauts already onboard, a procedure that became known as "load-and-go." The problem is that, from its very beginning, NASA had done things the other way. Hours before liftoff the launch team would load propellants onto a rocket before putting the vehicle in a stable configuration. Only then, with a fully fueled rocket, would a small group of pad technicians assist the crew in boarding.

The space station committee, chaired by revered former Apollo astronaut Thomas Stafford, felt that fueling a rocket with astronauts onboard was reckless. In 2015, Stafford wrote a letter to NASA saying as much: "There is a unanimous, and strong, feeling by the committee that scheduling the crew to be onboard the Dragon spacecraft prior to loading oxidizer into the rocket is contrary to booster safety criteria that has been in place for over fifty years, both in this country and internationally."

Later the Aerospace Safety Advisory Panel joined the chorus of concern. In its 2016 annual report, the panel said NASA must "scrutinize" the risks and hazards of this approach and warned that the space agency should not be unduly influenced by budget concerns of pressure to launch crew in a timely manner. These public remarks were measured compared to what was being said behind the scenes.

"When SpaceX came to us and said we want to load the crew first, and then the propellant, mushroom clouds went off in our safety community," McAlister said. "I mean, hair-on-fire stuff. It was just conventional wisdom that you load the propellant first and get it thermally stable. Fueling is a very dynamic operation. The vehicle is popping and hissing. The safety community was adamantly against this."

The advisory panels urged NASA to refuse SpaceX's request to employ load-and-go fueling, and the pressure ratcheted up after the AMOS-6 failure in September 2016, when the Falcon 9 rocket blew up during propellant loading. It was not difficult to imagine a similar, disastrous fate for astronauts inside a spacecraft. Although Crew Dragon had a SuperDraco-powered emergency escape system, pad aborts are dynamic and dangerous, subjecting astronauts to very high g-forces. The chilling video of the Falcon 9 being consumed by a massive ball of fire appeared devastating to SpaceX's cause.

Despite the safety community's insistence, however, Lueders did not tell SpaceX unequivocally no on load-and go. As she had done with propulsive landing, Lueders outlined the years of onerous testing and paperwork that would be required to meet NASA's safety requirements. But this time, Musk stuck to his guns. Densification formed a critical part of his plans to rapidly reuse the Falcon 9 rocket, and he would not yield.

True to Lueders's word, NASA made exacting demands of SpaceX to certify load-and-go for astronauts. Ultimately, what saved the company was its high flight rate. By the time the first astronauts launched on a Falcon 9, the rocket had flown more than fifty straight successful launches using load-and-go since the AMOS-6 failure.

"We tortured SpaceX for more than three years before we finally approved load-and-go," McAlister said. "We could easily have said no. There was a lot of pressure on us to say no and do things like we always had done since Apollo. Any other company would have given in. But this was critical to the reusability of Falcon 9, and it's a testament to Elon's single-minded vision."

Musk was determined that the Falcon 9 be reusable, and to make it economically viable the booster had to be loaded with super-chilled propellants. So committed was Musk that he willingly took on not just NASA but the entire human spaceflight community.

And he won.

"Without him," McAlister said, "there is no reusability revolution."

| 12 |

# A SECOND SPACE AGE

*April 20, 2019*

League City, Texas

On the Saturday before Easter, NASA astronaut Doug Hurley relaxed at his home near Houston after a long week of travel and training. This rare peace was soon shattered, however, when Hurley's mobile phone buzzed with a message from Lee Rosen at SpaceX:

"We lost the capsule."

Minutes earlier SpaceX had blown up the Dragon spacecraft, and Rosen wanted Hurley to know before the news came out. The vehicle, which Hurley could have been sitting inside instead of resting at home, was completely gone. Blown away.

It had all been going so well. During the first week of March 2019, the Crew Dragon spacecraft flew a stunningly successful mission to the International Space Station, known as "Demo-1." During the flight, which carried no astronauts, Dragon autonomously docked to the station for five days and completed a series of critical tests to demonstrate it could safely carry people.

Afterward, all that remained was for SpaceX to complete a few more parachute tests, and for NASA to complete a detailed analysis of data from the Demo-1 mission to clear Dragon for human spaceflights. During the week before Easter, Hurley had even visited the vehicle at Kennedy Space Center to size it up after the spaceflight.

As Hurley flew home to Houston for Easter weekend with his family, SpaceX prepared Dragon spacecraft for additional tests, including a firing of its SuperDraco thrusters. The company's technicians moved it to a stand at Landing Zone 1 ahead of the holiday weekend.

"We were on such a high," Hurley said. "After that Demo-1 mission, we knew we were going to do this. We were going to beat Boeing. We were going to space."

But then Hurley received Rosen's chilling text message. He told his wife, Karen Nyberg, a veteran of two spaceflights who had flown on the space shuttle and Russia's Soyuz spacecraft, what had happened. Pretty soon, he said, the disaster would be public. Together, they tried to process what this meant. Nyberg did not question whether Hurley should reconsider his commitment to flying on Dragon.

"Karen is not that person," Hurley said. "She's never going to say something like that. Instead, she asked me what I thought, and whether I was concerned."

These were questions Hurley could not immediately answer. For years he had been hearing the murmurings about SpaceX in the background, that the company was reckless and its actions were going to kill people—were going to kill him. But in the four years he had worked with SpaceX engineers, Hurley had seen another side. They were younger than NASA employees, certainly. But SpaceX engineers had passion and were deeply committed to safety and getting Crew Dragon right. He had bought in.

Hurley and Nyberg were still talking this through when, about an hour later, the story started breaking on the news. This was not an accident SpaceX could cover up, even if it wanted to. Large numbers of spectators had flocked to the gray sands of Cocoa Beach over Easter weekend for an annual surf contest, anticipating blue skies and white water. But that afternoon otherworldly, orange clouds of smoke began billowing ominously above the northern horizon. Soon, an unauthorized video emerged showing a countdown toward a firing of Dragon's SuperDraco thrusters. Then, shortly before ignition, the spacecraft violently exploded.

"That was the beginning of a few pretty dark months," Hurley said.

## Boeing has an astronaut problem

Douglas Gerald Hurley missed the era of internationally famous astronauts. He did not arrive at NASA until 2000, a time when the space agency regularly recruited large classes of fifteen to twenty astronauts for shuttle missions. There were so many of them, flying similar missions into low-Earth orbit, that shuttle-era astronauts lacked the celebrity of the Project Mercury pioneers or the Apollo Moonwalkers.

Even so, Hurley was cut from the cloth of the right-stuff era. With spiky blond hair and a chiseled face, Hurley had the looks and fighter pilot background to transcend the ranks of the mostly anonymous corps. He served as a Marine and a naval aviator before becoming a test pilot and flying advanced fighter aircraft. At NASA, one of his first assignments was working as a "Cape Crusader," watching out for the astronaut corps' interests in Florida. On the morning of February 1, 2003, he waited on the long runway at Kennedy Space Center for a vehicle that would never come. The loss of *Columbia* offered a stark reminder of the risks of his chosen profession.

After the shuttle returned flight, Hurley served as the pilot for two missions, including the program's historic final flight in 2011. Two years later, he stayed home with their son Jack as Nyberg spent six months on the International Space Station. After this mission, the couple considered their future. The only ride to space for the foreseeable future was onboard the Soyuz spacecraft, for a long stay at the space station. Hurley had no desire to do that. He had come to NASA to fly spacecraft, not be a caretaker and researcher and leave his family for months on end. When Brian Kelly, the director of flight operations at NASA, came to Hurley in early 2015 and asked if he was interested in flying, the pilot responded that only if it was something new.

He got his wish. In July of that year, he and three other astronauts, Bob Behnken, Eric Boe, and Suni Williams, were publicly named as NASA's Commercial Crew cadre. Eventually, the four astronauts would be split into pairs, with two assigned to Crew Dragon and two to Boeing's Starliner. But in those early years they often traveled together, visiting SpaceX and Boeing facilities to understand their spacecraft and provide input.

Hurley found the energy at SpaceX's factory in Hawthorne invigorating. "It was a stark contrast from what I was used to at Johnson Space Center, which was a somewhat sterile, stagnant, white collar, and older workforce. This was a bunch of kids, really. It was chaos, but a refreshing chaos."

Frequently, the crew would schedule a visit to Hawthorne with a detailed itinerary, but when they showed up a Falcon 9 launch date would slip, and it was all SpaceX hands on deck for that flight. The four astronauts would end up in a small room, catching up on paperwork or other activities. "They just didn't have the people to do a bunch of separate programs that were well into the future," Hurley said. "But you just had to roll with it."

Doug Hurley celebrates being named to the Demo-2 mission. | PHOTO CREDIT: NASA

Despite the disarray, however, Hurley preferred the SpaceX environment to that of Boeing. When the SpaceX engineers could be corralled, they were eager to hear feedback from the NASA astronauts, excited to work with them, and attentive to their suggestions. By contrast, Boeing engineers seemed indifferent to hearing from the four commercial crew astronauts. Hurley took this particularly hard because he believed it was due to a former NASA astronaut with whom he had been close.

Hurley had flown alongside Chris Ferguson, who commanded the final shuttle mission, and the two became friends. But their relationship had grown strained after the shuttle touched down. Just months after the flight, before the end of 2011, Ferguson left NASA to direct the "crew and mission" systems for Boeing's crew spaceflight program. Hurley believed Ferguson had plotted his exit from NASA to Boeing even before the shuttle flight. As part of this, Ferguson had carefully staged a flag ceremony during the mission, bringing to the space station an American flag that had been flown on the vehicle's first flight in 1981.

"This flag represents not just a symbol of our national pride and honor, but in this particular case, it represents a goal," Ferguson said as the flag was hung inside the station. "This flag will be flown prominently here by the forward hatch of Node 2, to be returned to Earth once again by an astronaut that launches on a U.S. vehicle, hopefully in just a few years."

Later, Hurley realized that Ferguson intended for that astronaut to be himself, flying on Boeing's new spacecraft. As NASA's Commercial Crew cadre started meeting with Boeing engineers, Hurley began to understand their detachment. Boeing already had *its* astronaut in Ferguson. He would command the vehicle's first crewed flight, and only his opinion really mattered.

"That was pretty off-putting for me," Hurley said. "I took it a little more personally because of the relationship that I had had with Chris. There was an arrogance with them that you certainly didn't see at SpaceX."

That confidence in Boeing's superiority was shared by the engineering teams at Johnson Space Center. They, too, believed Boeing would win the race. And who could blame them? NASA named its four commercial astronauts on July 9, a mere eleven days after the Falcon 9 rocket broke apart carrying the CRS-7 mission to the space station. Then, a little more than a year into the cadre's assignment, SpaceX blew up another rocket.

This marked a low moment for Hurley. While SpaceX wrestled with its Falcon 9 failures, Boeing also underperformed. Not only were its engineers overconfident, but the company's management also was not putting skin into the game. NASA had intended for its "commercial" programs like Crew to also attract private customers. Therefore, while NASA funded the lion's share of development costs, the companies were supposed to put money in, too. The idea was that this could be recouped from non-NASA missions. Yet Hurley did not see any urgency from Boeing's teams. Rather, they appeared to be working part-time on Starliner.

"It was all about managing the dollars and cents from Boeing's perspective," Hurley said. "And then on the other side, they're blowing up rockets left and right. It was just like, holy crap. I struggled pretty hard with that."

Yet in the months after the AMOS-6 failure, SpaceX started to win over Hurley and the other Commercial Crew astronauts. The company was transparent about its problems. Its engineers provided detailed updates on efforts to rebuild Space Launch Complex 40 in Florida and the Falcon 9's return to flight.

Meanwhile, his relationship with Boeing kept deteriorating. A breaking point came during the summer of 2018 as Boeing worked toward a pad abort test in White Sands, New Mexico. (Boeing never flew an in-flight abort test.) During a test-firing of the vehicle's abort motors in June, the engines burned normally. However, during the shutdown process a significant problem occurred due to a propellant leak. Ultimately, this would delay the company's pad abort test by more than a year, but at the time, Boeing neglected to tell the Commercial Crew astronauts about the issue. Hurley did not find out until nearly a month later and was incensed that NASA's astronauts had been left out of the loop.

"That was the straw that broke the camel's back for me," he said. "I was done with Boeing. And I don't think I did a great job of hiding it."

That summer NASA was closing in on making crew assignments for the first flights. Hurley told the chief of the astronaut office, Pat Forrester, he would not fly on Starliner. He did not want to fly again with Ferguson as the commander, feeling he had earned that right for himself. He also still fumed about the incident at White Sands being hidden from the crew. If that meant he was not going to space again, he could live with that.

The next month, during a ceremony with much fanfare at Johnson Space Center, NASA announced the crews that would fly on the Dragon and Starliner spacecraft. Hurley and Behnken were named to Dragon's mission. Ferguson would fly on the first Starliner mission, alongside Eric Boe and a novice flier, Nicole Mann. Suni Williams was named to Starliner's second flight.

As Hurley walked across the stage of the auditorium in Houston, he emphatically pumped his fists. The exuberance was genuine. He was thrilled with the assignment. At that moment it became personal for him and Behnken, who were great friends. Both were married to other NASA astronauts, had been each other's best man at those weddings, and had young sons. The race back to space began to feel like a sports competition, and they wanted to win. They agreed they were going to kick Boeing's ass. The next time the pair flew to Hawthorne for training, they told the Dragon team at SpaceX they were there to work for them. They were all in. Whatever it took.

## Leaky valves and asymmetric chutes

Two major failures nearly dashed those hopes. There was the Dragon explosion in April 2019, which became public almost immediately due to the ominous orange smoke clouds. But twelve days earlier another failure, also very serious, unfolded in Nevada out of public view. This accident occurred when a parachute failed and a Dragon "test sled" slammed into the ground and created a smoking hole. Solving both of these issues led to major delays.

For the Dragon explosion, Musk again turned to Koenigsmann, who was by then the longest-tenured employee at SpaceX. The German engineer had credibility both within SpaceX and at NASA and the other agencies investigating the failure. Inside SpaceX Koenigsmann was beloved as a fair, even-keeled elder statesman. And on the outside, regulatory agencies trusted him. SpaceX would need that reservoir of trust, as this was an ugly failure.

"It was a huge setback," Koenigsmann said. "The system that is supposed to save the crew instead blew up the capsule."

Koenigsmann worked alongside officials from NASA, the FAA, and the National Transportation Safety Board, but for a time the failure's root cause remained a mystery. It took Koenigsmann and his team about three months before they fully understood the issue, which involved nitrogen tetroxide. Most spacecraft, including Dragon, used this chemical as an oxidizer for its thrusters because it is storable at various temperatures for a long period of time, and it spontaneously combusts with hydrazine. Nothing in spaceflight is simple, but thrusters with this fuel mix are fairly straightforward: open the valves, introduce nitrogen tetroxide and hydrazine into the thruster, and presto, you're cooking with gas.

The anomaly occurred a mere 100 milliseconds before the SuperDraco thrusters were due to ignite. A leak allowed nitrogen tetroxide to enter some helium tubing, and when the propulsion system was pressurized a slug of this oxidizer was driven into a "check valve" meant to hold helium at bay. The titanium check valve subsequently ignited, precipitating the explosion of the spacecraft. This phenomenon was unexpected. Titanium had been used in spacecraft for decades and never caught fire like this. The solution was to replace the check valves with a different device, known as a burst disk, to regulate pressure. Because these disks do not open until they reach a certain pressure, they do not leak like a check valve.

Ultimately, the investigation led to a far better understanding of the behavior of tiny titanium metal shavings, under very high pressure, when mixed with an oxidizer. While this was a terrible accident, it provided critical insights into a failure mode of the SuperDraco thrusters. Kathy Lueders characterized Dragon's explosion as a "huge gift" because the failure had occurred on the ground, no one had been injured, and it advanced the understanding of a new phenomenon by NASA and the spaceflight community.

Though far less publicized, the parachute failure represented no less an existential crisis for Dragon. Initially, the spacecraft's engineers assumed parachutes would be a relatively straightforward problem. SpaceX had already landed multiple Cargo Dragons. And although Crew Dragon required four parachutes, due to the strain on the vehicle from added components relating to some abort scenarios, engineers believed going from three to four chutes would be a straightforward process. They were wrong.

Even though spacecraft have been landing beneath parachutes for more than six decades, the chutes remain a challenging problem. Imagine a flag blowing in a strong breeze. Its behavior is chaotic, flapping and waving in the wind. Parachutes are a bit like that, only on a much larger scale, with wind speeds of hundreds of miles per hour. Parachutes must deploy and inflate within a turbulent and dynamic airflow, encountering rapid changes in atmospheric pressure and changing wind speeds. Some of this can be modeled on computers, but there is no substitute for real-world testing.

What particularly troubled the Dragon parachute team was a variable known as an "asymmetry factor." This measures the chaotic nature of the four parachutes after their initial release, as they begin to inflate. Immediately after release the four main parachutes are crowded, jostling for space like four jellyfish crammed together. This jostling is never symmetric, and typically a single parachute will out-jostle the other three, or three parachutes will out-jostle one. The extent of this asymmetry is measurable, and in setting requirements from Dragon, NASA had given SpaceX a certain target. Above that, the agency said, parachute deployment could become too chaotic, with one or more lagging parachutes leading to serious problems.

Throughout much of 2017 and 2018, SpaceX began a series of two dozen tests of its "Mark 2" parachute design. This included dropping parachutes from different altitudes, in different conditions. As the parachute team got to work, they began to have all sorts of failures even though their design met NASA's asymmetry factor requirements. Essentially, they were finding that if one parachute was out-jostled by the other three and lagged too far behind, it would place too much strain on the fully inflated ones. The other three parachutes would snap, leading to a catastrophic failure.

"Imagine you find out deep, deep into your parachute development that the requirements given to you by NASA are nowhere near conservative enough," said Abhi Tripathi, whom Koenigsmann delegated to oversee parachute reliability.

By early 2019, after nearly completing an exhausting run of Mark 2 parachute tests, SpaceX and NASA realized they needed to beef up the parachute design to account for the higher asymmetry. This meant strengthening the parachute's canopy, and using a polymer material in body armor, Zylon, for the lines linking it to the spacecraft. Through the spring of 2019 SpaceX worked with its supplier, Airborne Systems, to produce the first "Mark 3" parachutes. Unfortunately, the new design meant SpaceX had to qualify the parachutes all over again and perform two dozen more tests. With increasing pressure to deliver Dragon, SpaceX moved into a rapid-fire test program. But on just the second test of the new Mark 3 design in April, while simulating a "one-out" failure to demonstrate that Dragon could land safely under three parachutes, the upgraded design failed disastrously.

"It was rapid-fire test number two," Tripathi said. "The key is the word rapid. But when you move that fast, the chance of having an anomaly is higher. With this test, with one chute out, a second chute failed shortly after, and then it cascaded. The whole thing slammed into the ground."

The subsequent investigation, led by Tripathi, found the problem was due to a device installed on the riser, part of the lines that connect the canopy to the spacecraft. The device, ironically, measured asymmetry, and it was covered with a protective cloth to prevent it from chafing the lines. (This cloth was actually a bed sheet "borrowed" from a hotel.) Video of the test, in which the parachutes were pushed out of an airplane at 10,000 feet, showed a white flash hitting the first of the three parachutes deploying. This was a strip of the sheet, traveling at high velocity.

That the fault was due to the test design, rather than the parachutes themselves, helped speed the analysis and get the SpaceX team back into testing during the summer of 2019. This team of technicians and engineers, led by Billy Burkey, an engineer who cut his teeth working on telescopes atop Mauna Kea in Hawaii, pushed madly during the second half of the year. They would take a step forward by completing a test, but then take two steps back after an anomaly or violating a test criterion, or some other problem.

The parachute team would drive to deserts in Southern California and Arizona, staging out of towns like El Centro, near the Salton Sea. They called themselves the "chute show," and they brought spouses and lovers into the desert because they spent days or weeks there at a time, building large bonfires at night, holding their own Burning Man–like bacchanals between tests.

"SpaceXers are kind of crazy workaholics," Tripathi said. "But the parachute team was a tribe unto itself. If you were on the parachute team, it was a factor of ten more. It was punishing and unrelenting, and these guys practically lived in the desert."

While the chute show busted its ass in the middle of nowhere, dropping parachutes and reviewing data, nearly all of the rest of the company turned its focus toward Starship during the summer and fall of 2019. Musk's prioritization of Starship—when Dragon was so close to being ready to fly, and the country so desperately needed its capability to break NASA's dependence on Russia for crew transport to the space station—pushed Jim Bridenstine to tweet about it being time for SpaceX to deliver. He was frustrated and angry, and he wanted to get Musk's attention.

I spoke with Musk in the immediate aftermath of the tweet. He'd staged the Starship event that had sparked Bridenstine's ire, revealing the first full-scale prototype of the vehicle in South Texas, on September 28, 2019. This had a nice resonance with history, as it marked the eleventh anniversary of the first successful Falcon 1 launch. Afterward, as the hour pushed midnight, *New York Times* science writer Kenneth Chang, a friend and colleague, and I were ushered into a small trailer at the South Texas factory. Musk waited inside, along with Paul Wooster, the company's Mars czar; and Claire Boucher, better known as Grimes, the Canadian singer and songwriter Musk was dating. We sat on upturned white buckets and talked about space for about thirty minutes. Bridenstine had clearly succeeded in getting Musk's attention.

"The NASA administrator was like, 'Why are you not working on Crew Dragon?'" Musk said. "Actually, there's nothing more we can do from a hardware standpoint. The constraint is mostly going through the

NASA safety reviews, and getting everyone at NASA, and SpaceX, and the FAA comfortable. This is like a zillion meetings. If you speed up the NASA reviews, we can launch sooner."

This was not entirely true. SpaceX still had parachute tests to complete, as well as an in-flight abort test of the SuperDraco thrusters. But Musk was accurate in saying that paperwork and reviews represented a majority of the tasks NASA and the company had to slog through before the crew launch could happen. Soon, he and Bridenstine reached a detente. A day or two after the Starship event, Musk called Bridenstine. Despite the ruffled feathers, the chief of NASA and the leader of SpaceX worked through their frustrations. Two weeks later Bridenstine visited the SpaceX factory in Hawthorne, and Musk updated him on the company's efforts to develop and test the new Mark 3 parachutes, along with other Dragon efforts. "Elon has told me, and he's shown me that that's where their priority is," Bridenstine said during the visit. "They're putting as much resources and manpower as they can to getting those parachutes ready."

During Bridenstine's tour, Musk told reporters that Dragon was "absolutely the overwhelming priority" for SpaceX. Internally, he had already informed employees as much. Laura Crabtree, who had been training Hurley and Behnken for their flight, said the atmosphere at SpaceX changed overnight. "It was like this light bulb went off," she said.

There have always been competing priorities at SpaceX. A decade earlier Cargo Dragon competed with the Falcon 9 rocket for resources. Then came the Grasshopper and recovery programs, followed by Starlink and Starship. There were only so many engineers, programmers, and technicians to go around, and always more work to be done. For Crabtree it could be frustrating when she tried to schedule time to meet with engineers responsible for various parts of Dragon or to get software work done.

But after Bridenstine said it was time to deliver, and Musk responded that SpaceX would deliver, all those barriers fell away. When Crabtree or another member of the Dragon team asked for something, they no longer had to wait. When more people were needed to throw at a problem, those resources were given. Flight software was delivered on an accelerated schedule.

As the year 2019 turned into 2020, the parachute team had nearly run through its two dozen tests, and NASA was satisfied with the SuperDraco fixes. The space agency had most of the data it needed. After a successful in-flight abort test of Dragon in January, putting both the parachutes and SuperDraco thrusters to the test, it was mostly a matter of finalizing the necessary paperwork and slotting the crewed mission into the space station's schedule of visiting vehicles.

By mid-April, even as COVID-19 lockdowns were becoming increasingly widespread, NASA was confident enough in Crew Dragon to set a launch date for the mission. Hurley would command, with Behnken overseeing operations. They were due to launch on the afternoon of May 27.

## Musk has questions about the weather

The weather sucked.

Due to the orbital dynamics of reaching the International Space Station, there typically is one opportunity per day to lift off from Kennedy Space Center. Essentially, the rocket launches as the station passes overhead, and a spacecraft spends about a day raising its orbit and catching up. For late May, this meant Dragon needed to launch during the late afternoon, at 4:32 PM. Unfortunately, the summertime pattern along Florida's Atlantic coast often sees thunderstorms fire up along the sea breeze during the afternoon hours.

That's what happened on Wednesday, May 27. About thirty-five minutes before the planned liftoff, SpaceX's launch director Mike Taylor made the call to proceed with loading liquid oxygen onto the Falcon 9 rocket. He knew it was touch and go with the weather but thought there might be a chance to get off the pad.

At the time the decision was made, Musk wanted to know who ultimately was responsible for making the call on weather. Dressed in a black suit, Musk had stationed himself inside Firing Room 4 at Kennedy Space Center. For its Crew Dragon missions, SpaceX decided to operate the launch of the rocket from this NASA facility, instead of its launch control center a few miles away, on the Cape Canaveral side of the spaceport. Cameras were there to capture footage of Musk for a documentary titled *Return to Space.*

"I just want to double-check, who is doing the measurement?" Musk asked.

"The Range," replied Kiko Dontchev, who was the director of Dragon ground and launch operations for SpaceX.

"Yeah, great, but who at the Range?"

"Uh, the Launch Weather Officer."

Musk was fishing for someone to hold accountable in case the Falcon 9 rocket scrubbed. Like almost everyone in the launch control center during that first attempt, his nerves were fraying. The officer in question was Mike McAleenan, who was closely monitoring weather conditions in a Range building a few miles away. McAleenan was not optimistic as a strong thunderstorm rolled through about half an hour before the instantaneous launch window. He even had to sound tornado sirens at the Cape as a twister touched down about three miles away from the launch pad. But not to worry, McAleenan told the SpaceX launch center, the tornado was moving away from the rocket.

On top of the Falcon 9, Hurley and Behnken waited anxiously. There were just two windows in Dragon, near the entry hatch. From their seats, this made it difficult to see weather conditions outside unless it got really dark. As the clock counted down, they could discern blackening skies and raindrops on the window. Hours before boarding Dragon, they had known the weather would be tight. Hurley always harbored doubts about launching on time, so scarred was he by his first spaceflight in the summer of 2009. That mission scrubbed twice in June due to a hydrogen fuel leak with space shuttle *Endeavour*. After that problem was fixed, early evening launch attempts were scrubbed on three consecutive days in July. Only on the mission's sixth attempt did Hurley get his first ride into space.

"Because of that, whenever I go to the pad, I assume we're going to scrub," he said.

McAleenan spoke frequently with Taylor after liquid oxygen loading began. While the storm had been gnarly, it was starting to clear. He forecast that conditions would improve by the time the rocket was supposed to take off. But there was a problem with the "field mill rule." This is an arcane-sounding term that essentially means residual electricity in the air after a thunderstorm moves through. The atmosphere gets charged up by lightning, and it takes time for the electric field to settle down. McAleenan said he needed another five or ten minutes added to the launch window so that field levels would reach acceptable limits.

But Dragon did not have more time. Once SpaceX committed to fueling the rocket, it had to go precisely on schedule. And the instantaneous launch window for the space station also meant that, if Dragon launched late, it would not have enough propellant to catch up to the station. With seventeen minutes left before liftoff, Taylor called a scrub.

In the SpaceX firing room, Musk rolled his eyes in exasperation. Looking outside he said, "At least eyeballing it, this does not look like we should be stopping launch." Speaking to Dontchev, he added, "We just wanna make sure we're not scrubbing based on weather that's half-hour-old. We just wanna make sure it's real and we're not jumping at shadows."

With all of the weather issues and intense public scrutiny, it had been a tough day for McAleenan and Taylor, close friends after working on dozens of previous launches. They agreed to meet at Nolan's Irish Pub in Cocoa Beach that evening to drown their sorrows. However, as McAleenan drove to the pub, Taylor telephoned him. The SpaceX launch director warned McAleenan that he might get a call from a strange number in a few minutes, and that he probably should answer the phone. It would be Musk. McAleenan turned around and headed for home, organizing his Range documents and materials so he could explain in detail why the launch had scrubbed.

Sure enough, Musk called. He listened as McAleenan explained how thirty-three coffee-can-sized field mill devices are spread across Cape Canaveral, each measuring volts per meter. They are very sensitive to the point of recording a spike when a lawnmower passes by, picking up on charged particulate matter. When the electric field is elevated by a thunderstorm, the Range rule is to wait fifteen minutes for the atmosphere to clear.

Why was this necessary? Because when a rocket launches it punches into the air and compresses the electric field. Even inside a cumulus cloud not generating lightning, this compression can combine with a rocket's ionized plume to trigger a lightning strike. Famously, this happened during the Apollo 12 mission to the Moon. The massive rocket triggered two strikes less than a minute after lifting off, and mission control had to make a rapid-fire call on whether to abort or continue the flight. Ultimately the crew made it safely to space after resetting some of the electronics inside the Apollo spacecraft. After this, NASA and the Range began to develop field mill rules.

As Musk listened, he began to appreciate the rules. Yes, they probably were a little too conservative, but they were grounded in physics. He asked McAleenan to email him a copy of the rules documents, which Musk said he intended to read later that night. After hanging up, McAleenan took a deep breath and went to the pub.

"If you give Elon the straight answer, with all of the detail he wants, he's very happy," McAleenan said. "But don't try and bullshit him, because he'll smell that out."

Another observer of the first Dragon launch attempt was less swayed by explanations. It was an election year, and Florida was an important swing state. President Trump liked the symbolism of America returning to space under his watch, especially since the shuttle had stopped flying during Obama's tenure. So Trump wanted to be on hand when NASA, in his eyes, became great again. He watched the countdown from a balcony on the Operations Support Building II, which overlooks the launch pad from a few miles away. When the launch scrubbed, Trump was furious. He did not understand why, and really did not want to know. Instead, he blasted Bridenstine in front of the crowd of VIPs assembled for the launch.

Three days later, on the next available launch date, Hurley and Behnken again donned their SpaceX suits and rode in Teslas to the launch pad. This time as they took the elevator to the top of the launch tower and boarded Crew Dragon, they felt confident. The weather was better, and from a technical standpoint the first countdown had gone smoothly.

Inside Dragon, Hurley listened to the noise of SuperDraco valves opening and closing. The rapid-fire checks sounded like a rapid hammering. Farther below him, Hurley could feel the rocket being loaded with propellant. He spent much of these minutes in reflection. Hurley believed in the SpaceX and NASA teams that had placed him here, at the point of the spear. Since 1961 only eight different human spacecraft had flown into orbit, all built by the Soviet Union, United States, or China. The last debut of a new crew vehicle, China's Shenzhou spacecraft, had been nearly two decades earlier. It was both a privilege and liability for Hurley and Behnken to be sitting in these seats. Hurley also thought about Karen and their son, Jack, watching the venting rocket from a few miles away. They were nervous, and so he worried about them, worrying about him.

Yet mostly Hurley thought this just needed to happen. NASA had been without this capability for nine years, and America needed to get back into the human spaceflight game. "It was time," he said. "It was 2020, the middle of a pandemic, and the team had worked so freaking hard for so many years."

And then it *was* time. At 3:22 PM locally, the Falcon 9 rocket leaped off the pad and continued accelerating toward space against a bright blue backdrop. The ride felt much smoother than the space shuttle because that vehicle had extremely powerful solid rocket motors that rumbled and roared and provided most of the kick to reach orbit. Hurley could hear the wind outside, whistling around Dragon. Then came the eerie feeling of the first stage shutting down, and the pneumatic separation of the stage. This threw the crew forward in their seats. After a few seconds, the second stage engine lit. This part of the ride was bumpier because the engine was so close by, only a matter of feet behind the spacecraft.

As the upper-stage Merlin burned, Hurley turned to Behnken and said, "It's riding a little rougher than I thought."

But Dragon was riding true.

## Small town guys change world, eat pizza, drink beer

Laura Crabtree should have been sleeping, as she was due to work the overnight shift at mission control in Hawthorne. But she had come to

SpaceX in 2009 to put people into space, and she had trained Hurley and Behnken for the last five years. They were her friends, and no way was she going to sleep during the most dynamic part of the mission.

Coming into the Hawthorne factory on launch day felt strange. The city streets surrounding SpaceX were eerie, and the mood tense. Protests in Los Angeles over the murder of an African American man, George Floyd, were reaching their peak. Crabtree and the other employees weren't sure whether the roads would be closed in Hawthorne when they tried to go home after the launch. The background of the COVID-19 pandemic further amplified the unease. SpaceX had an exemption to keep operating with on-site employees, but limits were placed on gatherings. During launches a crowd typically gathered outside the Mission Control Center, watching proceedings through floor-to-ceiling glass windows. But due to the virus, no one could congregate there now.

"It was terrible," Crabtree said. "We were doing this crazy thing, and traditionally you've had thousands of people cheering you on. And it really pumps you up. But this time it was crickets. Maybe someone mopping the floor."

She and a cohort of the operations team banded together in their office area on the third floor of the building, above mission control. Crabtree followed along, listening to the various loops of flight controller discussions. Because of the earlier scrub, she was not sure Bob and Doug would launch this time. It was not until the final minute that she realized they were actually going. Her heart began to pound and flutter.

David Giger watched the launch away from the factory, at his home in Santa Monica. He'd left SpaceX at the end of 2017, after more than a dozen years. SpaceX and Dragon had become a large part of his identity. But Giger was no longer a kid in his twenties, and the grind of working eighty to a hundred hours a week no longer appealed to him. He married a woman named Arleen, a technical recruiter at SpaceX. They did not want to postpone their lives any longer, so they left and started a family. Giger watched the Demo-2 mission with his two-month-old daughter, Ava, on his lap.

The couple had mixed emotions. They missed out on the thrill of the launch but also could savor its success. Giger had built the team that built Dragon. His wife had recruited many of the engineers who worked on it over the years.

"I felt so proud of the awesome machine we built together," Giger said. "It was a long, hard fight with the best team in the business."

Before she left for Kennedy Space Center on the morning of the Demo-2 launch, Kathy Lueders told her husband not to expect her home that evening. She packed a small overnight bag and brought a pillow and blanket to the firing room. Lueders observed the launch from a station near Koenigsmann, and after the successful liftoff they embraced.

Together, they had worked two rocket failure investigations, and then the Dragon explosion. Like Hurley, Lueders had welcomed the radical transparency SpaceX offered. When there was an issue with the rocket or spacecraft, Koenigsmann or his chief deputy, Bala Ramamurthy, let her know immediately.

"Every time there was a problem they would call," she said. "They didn't hide anything. They'd say, 'Let's go work it together.' That's partnership. That's been the important part of the relationship. Spaceflight is very, very, very hard. I don't think we can say it enough."

As leader of the Commercial Crew program, she felt as though the crew's lives were in her hands. She had signed off on the safety of Crew Dragon and the load-and-go launch procedures for the Falcon 9 rocket. Many of her colleagues at NASA questioned this. But she believed in what SpaceX was trying to do. The most vulnerable part of the mission for the crew was the launch and flight to the station. For the Demo-2 mission, this took about nineteen hours. Lueders wanted to be on hand, with the families of the crew, in case something went wrong.

So she spent the night at Kennedy Space Center, trying to sleep on a small couch in a break room. But people kept waking her up as they came to get coffee. A SpaceX employee took pity on her and invited Lueders to crash in a small conference room.

The two astronauts made it to the station and spent two months living onboard, with Dragon checking out fine during its extended stay in orbit. Then it was time to come home, making the final and perhaps most hazardous part of the journey. Of all the risks faced by a spacecraft coming back to Earth, three stand out. The first is the loss of communications during a period when the plasma streaming around the spacecraft is super-hot. This blackout period lasts for several minutes after a vehicle enters the atmosphere and is nerve-wracking for ground controllers who lose touch with the people and spacecraft.

SpaceX also had concerns about the asymmetric backside of Dragon. This is due to the four engine pods, each containing two SuperDraco thrusters, that stick out from its sides. This lack of uniformity in its shape could cause Dragon to roll as it traveled at hypersonic speeds through the atmosphere. Hurley said he and Behnken were hyper-focused during that time, but Dragon flew just fine.

And finally, there were the parachutes.

Abhi Tripathi followed reentry from the third floor of Hawthorne, alongside the twenty members of the parachute team clustered around Billy Burkey's cubicle. Tripathi had spent a year embedded with the chute show, which had performed test after test, astonishing NASA with their speed. But not all of the tests were successful. There had been more than one test failure in the desert.

"I was very, very nervous," Tripathi said. "We had had multiple anomalies of multiple types. I watched as the drogues deployed, and the mains deployed, and then finally splashdown. That's when the weight came off my shoulders."

Seeing the big, billowing, white parachutes carry Dragon softly down into the Gulf of Mexico also broke Crabtree's composure.

"It was just full-on tears as the parachutes deployed," she said. "They were tears of relief and tears of joy. It was every emotion, all at once. Most of us had worked five years, or ten years, to make this happen. This was the one thing I wanted to do, to take people to space. We had accomplished something only nations had done, and during such a dark period of life around the world."

As Dragon floated in the water, Hurley and Behnken had one final objective to complete before they were fished out of the sea and the hatch opened. NASA had provided a satellite phone to be used in case Dragon landed far off course and they lost radio communications. There were a couple of people the crew were supposed to call to verify the phone worked, and the numbers were programmed into the phone. But when they called the NASA flight surgeon in Hawthorne, he didn't answer. Similarly, a call to an official in Houston went unanswered. So they broke the script and called their wives, Karen and Megan, in mission control at Johnson Space Center. Hurley also called his mom. To kill time, they dialed other friends whose numbers they could remember.

They were just so happy, bobbing in the ocean. Deliriously happy. They'd done it. All the years, all the work, all the setbacks, and it ended in success for them, for SpaceX, for NASA, and for the United States of America. Hurley and Behnken were heroes. Dragon was the first new U.S. spacecraft to fly astronauts into orbit since the space shuttle's debut four decades earlier. And *they* had taken the ride, the risk, and reopened the nation's highway to space. They were helped aboard a helicopter from the recovery ship to the shore, and from there onto a NASA plane back to Houston and their waiting families.

Before the landing, one of SpaceX's astronaut liaisons, Haley Esparza, had asked the crew if they desired anything for the plane ride. Hurley said he wanted Fat Tire amber ale and some pizza. And that's exactly what awaited them onboard. It was an All-American ending. Just two buddies from small towns, eating pizza and drinking beer after flying inside a fireball back to the planet. They'd captured the flag and brought it home. And now they were going home, too, having changed the world forever.

"It was the second space age," Hurley said. "And it started in 2020."

| 13 |

# STEAMROLLER

*February 5, 2018*
Kennedy Space Center, Florida

A black luxury sport utility vehicle rolled slowly along a single-lane, paved road just inside the Launch Complex 39A fence line. It drove directly up a ramp, onto the launch pad. When it stopped, Elon Musk and his five children tumbled out, walking right up to the base of the titanic Falcon Heavy rocket. Sunlight glinted off its three lambent white boosters.

Nothing like this beast had flown successfully before. During the era of America's triumphant Moon landings, the Soviets countered by trying to launch a massive rocket of their own. This machine possessed an almost unfathomable thirty first-stage engines. Four times the brilliant Soviet rocket scientists launched the massive N1 booster, and four times it failed, creating some of the most spectacular explosions ever seen. But the rocket never got close to the Moon. There were too many engines to manage.

Now, in early February 2018, SpaceX would try where the Soviet Union failed half a century earlier. After Musk's family completed their tour, the SUV drove around the ring road to a vantage point a quarter of a mile from the launch pad. After a few minutes, Musk emerged for a pre-launch interview. He straightened his black sports jacket and then smiled impishly as he walked over and shook my hand. "It's small, don't you think?" Musk asked. "I think we need to step up our game."

I was not sure what to think. The rocket was not small in any way. It looked suitably massive and imposing. Later, I would learn this was not in fact a joke. Musk had just surveyed the largest, most powerful, and most capable rocket in existence. It would blast off the launch pad the very next day, with the energy of 4 million pounds of TNT. For the entirety of its sixteen years, SpaceX had strived toward this, buffeted by headwinds on all sides most of the way. The history of this moment, and this pad, was palpable.

Yet Musk's first reaction was essentially, "Meh."

Why? To know the answer is to understand Musk. He lives with one eye fixed on the present and the other looking firmly into the future, where he was already deep into the design and early development of the Starship rocket. And as big and bad and brawny as the Falcon Heavy might be, it was merely a stepping stone.

But oh, what a stepping stone it was.

## "Nothing felt unachievable."

Long before I met Musk on the eve of the Falcon Heavy debut, even before the Falcon 9 took flight, he began plotting development of a heavy lift rocket. Musk and the propulsion team designed a vehicle that combined three Falcon 9 cores into a single first stage. This unsettled engineers like Kevin Miller.

"I was living with trying to go from one engine to nine," Miller said. "I couldn't even imagine reliably igniting nine engines at the time. I wouldn't say I was afraid of it, but I knew that building something like the Falcon Heavy was going to be extremely difficult."

Tim Buzza created a graphic that depicted an early version of the rocket launching from Cape Canaveral. For fun, he sent a copy to Miller, quipping that the propulsion engineer needed to step up his game to ignite and control twenty-seven engines at a time. As a joke—well, sort of a joke—Miller printed the graphic and wrote "Retire before this happens" on it. He placed this on his desk as a reminder. Miller did not make it, remaining as the company's principal propulsion engineer when the Heavy flew.

Kevin Miller's note to himself regarding the Falcon Heavy. | PHOTO CREDIT: KEVIN MILLER

Musk spoke publicly about the big rocket for the first time in the spring of 2011. The venerable space shuttle would retire in a few months, but Musk had good news for space fans. The Falcon Heavy rocket could lift 117,000 pounds to orbit, twice the capacity of the shuttle. Putting this into perspective during a news conference at the National Press Club, Musk explained this equaled a fueled Boeing 737, with passengers in every seat and a full luggage compartment. "This is a rocket of truly huge scale," Musk said. "This is something America can be very proud of."

He was asked when the Falcon Heavy would make its debut launch. It should be ready to launch at the "end of next year," Musk said. This was a quintessential green lights-to-Malibu prediction. The rocket didn't exist. Its launch pad didn't exist. And SpaceX engineers were up to their eyeballs building Falcon 9s and Dragons.

SpaceX would miss Musk's target by more than five years. Certainly, they could have built the rocket sooner, but it lost out to competing priorities. NASA was paying for cargo missions on the Falcon 9. Moreover, as the smaller rocket got more powerful, customers who had signed up for the Heavy at a later date could be moved to the Falcon 9. Most critically, SpaceX needed to find half a billion dollars to develop the Heavy and time for its employees to work on it.

The company's priorities began to change in the mid-2010s, as SpaceX forced its way into the bidding for military contracts against United Launch Alliance. The Air Force and its spy agencies have nine "reference" orbits that, as a matter of national defense, it has deemed necessary to reach. Only the Delta IV Heavy, manufactured by United Launch Alliance, could hit all of these orbits with large payloads. If SpaceX wanted to be on par with this competitor for the full range of military launch contracts, it needed a big rocket, too.

A major challenge in designing a rocket with three cores is understanding the interaction of these powerful boosters with one another. The noise and energy generated at liftoff is three times greater than a single Falcon 9 launch, some of which is channeled sideways from core to core. SpaceX engineers spent a lot of time studying these interactions, concerned that an unexpected resonance might lead to a structural failure.

"I was concerned about booster-to-booster interaction," Musk said. "You've got a lot of dynamics going on there. Those rockets are very flexible, and if they flex in unexpected ways, they could potentially impact one another."

Some of this work fell on Chris Hansen, the director of testing. His team built special stands in Hawthorne and McGregor to test the structural integrity of the triple-core rocket, the ability of the side-mounted boosters to separate cleanly in flight, and the joints connecting the boosters. His already overworked team of engineers and technicians received another task in the fall of 2017, when Musk revealed his intention to launch his midnight cherry Tesla Roadster on the rocket's first flight.

Several of Hansen's technicians had come to SpaceX from jobs where they customized off-road vehicles. Musk asked them to harden his Tesla for the rigors of spaceflight, as cars are not built to withstand high g-forces or the vibrations of launch, not to mention a vacuum. Methodically, Hansen's team worked through each component, adding structural support to the side panels so they would not dent, and removing the suspension and replacing it with a rigid structure. They did this behind a curtain, in the corner of the factory, to maintain secrecy.

Musk did not announce the payload until December of that year. He received criticism for this, with outsiders saying it was shameless cross-promotion of Tesla. For a time, employees like Hansen wondered, too, why they were working hard to launch a whimsical payload like an automobile. Musk had actually offered NASA a free ride on the rocket for one of its missions, and the space agency had declined. The Tesla was a backup plan.

"As my guys worked on the car, I thought, *Is that really what we're going to go launch?*" Hansen said. "But then I had a 180-degree change in my viewpoint once I saw the fairing separation." The revelation of that Tesla, with its spacesuit-clad "Starman" mannequin in the driver's seat and Earth in the background, was an arresting visual. "I understood why Elon was doing it. It made space fun. It was inspiring."

During the run-up to the rocket's launch in February 2018, an accident in Texas nearly delayed the mission by several months. Engineers were running a thrust-vector control test on the rocket's center core and ran out of hydraulic fluid to steer the engines. The bell-shaped nozzles went haywire, slamming into one another, and the incident became known as Hell's Bells. This was the very nightmare scenario that had kept Miller up late at night a decade earlier. The resulting contorted nozzles resembled the swirling human figure in Edvard Munch's famous painting *The Scream*.

SpaceX did not have nine spare engines. A majority of missions at the time still used brand-new first stages, pushing the Merlin production team to its limits. Replacing the engines would also add millions of dollars to the cost, which SpaceX was paying out of pocket. To solve the problem, Musk called an old friend, Marty Anderson. The technician with steady hands and tin snips who saved the day back in 2010, on the second launch of the Falcon 9, had retired a few years earlier. This time there were nine Merlin engine nozzles messed up, not one. Anderson leapt at the challenge.

He built a rounding ring and set to work with clamps, leather, and a set of hammers. He could not bang on the nozzles, because the metal would smash. However, by gently tapping the nozzles with a hammer, Anderson could coax them into position. Over the course of five weeks, he worked meticulously, like a blacksmith of old, rounding the engine nozzles back into shape. By the time he was done there was not a dent or perceivable ding.

"Elon and Gwynne had been so good to me, and all of those people were my family," Anderson said. "So it was really important for me to get those engines fixed." One of the reasons Anderson retired is that he felt, as the company's manufacturing processes matured, his maverick skills were of less use. But now, instead of just fading away, he could go out with a bang. "I went out doing the greatest thing I had ever done," he said.

Anderson did save the day. The engines were shipped back to Texas, and from there to Florida. Three days after Christmas, in 2017, the three united cores rolled to the launch site for the first time. Everything about integrating the rocket proved more challenging than a single stick booster. For example, when technicians plugged the Falcon Heavy rocket into the launch pad at Kennedy Space Center for the first time, the volume of telemetry data choked the launch pad's ground software.

In the four weeks after rollout, several fueling tests were performed to ensure the ground systems could provide enough densified liquid oxygen for three boosters. During these tanking tests, the structures team also monitored the twisting forces the side boosters applied to the central core, with dozens of sensors up and down the rocket. It took several fueling attempts for the team to gain confidence in their measurements and finally complete a static fire test at the end of January 2018.

Although the payload was just Musk's old car, SpaceX had a lot riding on a successful liftoff. Final modifications were underway on Launch Complex 39A to accommodate human missions, including installation of a sleek crew arm for astronauts to board Dragon. Moreover, SpaceX had only brought SLC-40, so badly damaged during the AMOS-6 failure, back online six weeks earlier. As it sought to ramp up its flight rate, SpaceX could ill afford to destroy its crew launch tower.

Although Ricky Lim had presided over dozens of launches before, his gut twisted during the countdown on February 6. The upper-level winds were bad, and this delayed the planned liftoff time by about two more hours. And he worried a lot about losing the pad.

"As many launches and countdowns as I can recall, I literally remember my heart beating out of my chest with Heavy," he said. "Because we knew that if it blew up, or something really bad happened, it would also put the crew tower at risk. That was very nerve-wracking."

When the winds finally dropped to acceptable limits, those concerns evaporated. Twenty-seven Merlins roared to life. The rocket ascended majestically from the launch pad where the Saturn V had once carried astronauts to the Moon. Among the awed bystanders were Miller and his mentor, Tom Mueller. They watched from a vantage point between the launch and landing pads, two proud papas, as twenty-seven Merlin engines produced a stunningly bright flame. They were speechless. They had struggled and sweated and despaired with a single engine for so long. Now there was this magical monstrosity.

Soon, the rocket was gone—only not entirely. Two dots appeared in the sky, steadily brightening as the side boosters returned to their landing pads on the coast. Seconds after the boosters lit for their landing burns, they disappeared below the tree line. As this happened, the sound of the Merlins washed over the onlookers. Could anything be better?

"I don't think people really understand how fast a rocket is going," Mueller said. "It takes off, and it's gone, supersonic. It's this giant thing, and it's just gone. And then it's not that long later and you see these two lights way the hell up in the sky. And you can see the white streak coming in, and it just seemed way too fast. You're thinking, *It's not going to stop in time*. The forces and the velocities involved are just incredible. It was literally fiction until we did it. So you were watching science fiction. I thought it was astounding."

So did all who saw it that day. It is difficult to overstate the effect of the launch, the iconic flight of the Starman mannequin in a Tesla, and that dual booster landing on the general public, the space industry, and on SpaceX itself. In that moment, the present overtook the future.

Before the mission, some critics labeled the Falcon Heavy a vanity project. It was just the Tesla guy using his other company to build a giant toy. Serious people in the industry told me this. But in truth, the rocket has been a boon to the U.S. government. When NASA needed a rocket for arguably its most ambitious science mission launch of the 2020s, a spacecraft that will make dozens of flybys of Jupiter's moon Europa, it had two choices. The first was using its own Space Launch System rocket; the second was the Falcon Heavy. By NASA's own estimate, its rocket costs $2.3 billion per launch. Instead, NASA signed a contract to launch the Europa Clipper on the Falcon Heavy for $178 million, saving taxpayers more than $2 billion.

The entry of SpaceX into the competition for national security launches also brought down costs for the U.S. military, particularly for large satellites required by the National Reconnaissance Office. "We've saved 50 percent over what the older prices were like for the Delta Heavy," said Colonel Douglas Pentecost of the U.S. Air Force. "We are just saving a ton of money on the high-end." At the end of 2023, with the imminent retirement of United Launch Alliance's Delta Heavy rocket and its next-generation Vulcan rocket still awaiting certification, the Falcon Heavy is the military's only lifeline to get its largest and most valuable payloads into orbit.

The Falcon Heavy had a still more profound effect on SpaceX employees, while inspiring the young engineers who would come to work on the Starship project. It marked the culmination of everything SpaceX had done during its first decade and a half. The company built a rocket with one engine, then nine, and now three times that. During just the previous year, SpaceX roared back from its second failure with eighteen successful launches. It proved that landings worked and that a private company could build the world's largest rocket. The flawless liftoff and dual booster landing hinted that the audacious Starship project might be possible after all. It marked the end of the beginning for SpaceX, and the beginning of what would come next.

"Nothing felt unachievable at that point," Hansen said.

## What the first landed Falcon 9 really looked like

After the ORBCOMM launch and landing just before Christmas in 2015, Musk was eager to learn whether the first stage might be reflyable. For the first time he had in his hands actual rocket hardware that had flown to space and returned. He told John Muratore and the rest of the launch team to put it back on the pad and test-fire it within a week. Muratore did so, but with some trepidation because the rocket's condition was unknown.

"Elon always had the guts to try something that nobody had ever tried before," Muratore said. "But that was a pretty big risk, because we were worried it would take the pad down."

Still, they pressed ahead. Following a cursory examination, the launch team ignited the first stage. All nine engines lit successfully, but then one shut down prematurely, followed soon by the other eight. An examination of the interior of the rocket uncovered the problem. The liquid oxygen tank sits atop the kerosene fuel tank, and a tube carries oxidizer down through the fuel tank to the engines. The fiberglass insulation that coated this line had completely blown out, sending foam throughout the kerosene tank.

Though happy with the test, Musk also felt disappointed. He would have liked a full-duration burn to show the world the Falcon 9 could soon fly again. But this rocket would not do so. First the company shipped it to Texas for additional tests, and then Musk wanted the rocket preserved as a trophy. Today, it stands outside the company's headquarters, on permanent display for the world to see.

The company's second rocket to land, however, did fly again. After the CRS-8 rocket touched down on a drone ship, the company towed it back to shore and spent more than a year disassembling and inspecting the booster. Every engine was taken apart and updated with a new internal fuel pump component. The engines subsequently underwent thorough testing in Texas.

Many parts of the rocket were swapped out, wholesale. The interstage that sits on top of the first stage was new, as were nearly all of the avionics boxes. A lot of the main propulsion components, such as valves and regulators, were replaced. The majority of the cork insulation at the bottom of the rocket was repaired, a truly tedious task that involved stripping, trimming, and gluing new sections of the material. A lot of this had to be done by hand. Finally, the booster was cleaned. SpaceX has since dispensed with this last process, leaving used rockets in their sooty state. But for the first reflight mission, an army of technicians and engineers used gallons of isopropyl alcohol and rags to scrub the vehicle until it had a brilliant white sheen once again.

As this work proceeded, this booster also needed a customer willing to risk a payload. "It wasn't like we had Starlink missions where we could just go and fly them, and if we had a failure it would be painful, but it wouldn't be the end of the world," Zach Dunn said. "The stakes were very high, with a relatively small number of missions at the time and high-dollar customers."

SpaceX did find a willing partner in the Luxembourg-based satellite company SES. Three years earlier, the European officials had been ready to ship their SES-8 satellite back across the Atlantic before Muratore assured them that the SpaceX "maniacs" could put the launch pad together quickly. The Luxembourgers liked what they saw and launched another satellite with SpaceX in 2016. Now they had the SES-10 satellite ready to deliver telecommunications services to Latin America. Their experiences with the launch team, and the rocket itself, gave the satellite company confidence.

As an inducement, Shotwell offered a 10 percent discount from the Falcon 9 list price of $62 million. This helped, of course. But SES had shown a proclivity for taking risks to help push the industry forward. The launch of their SES-8 satellite had been the first time SpaceX boosted a payload to geostationary transfer orbit. Now they would take on reuse.

"We did receive a discount," said Martin Halliwell, chief technology officer for SES, just prior to the launch. "But it is not just the money in this particular case. It's really let's get this proof-of-concept moving. Someone has to go first here, and SES has a long history of doing this."

Ricky Lim again had responsibility as launch director for the mission as it rolled out to the launch pad in March 2017. The company had inspected, tested, and serviced every part of the rocket they could think of. It should work. But would it? Another failure just half a year after AMOS-6 would have greatly shaken the Falcon 9's long line of customers. "We felt as prepared as we could, but extremely anxious," Lim said. "Attempting such an audacious feat the year after two major anomalies, we were understandably paranoid."

On a clear evening in late March, the first previously flown Falcon 9 booster took flight from Launch Complex 39A, just the company's fourth mission since the AMOS-6 disaster. After a flawless launch, as a cherry on top, SpaceX recovered a payload fairing intact for the first time.

SpaceX had demonstrated the fundamentals of rocket reuse. But the first stages required extensive maintenance and months of careful testing before reaching the launch pad again. None of the "Full Thrust" versions of the Falcon 9 rocket ever flew more than three missions. Technically, this counted as reuse, but it remained far from practical. To break even financially, given the mass penalty from carrying fuel for landing, refurbishment costs, and financial investments in recovery, SpaceX needed to fly each first stage at least several times. And Musk knew that to unlock a future of rapidly reusable rockets, his first stages must be capable of many flights, with minimal refurbishment.

His engineers spent most of 2016 and 2017 breaking apart returning rockets to better understand how to go about that. They put all of these learnings into one final upgrade to the Falcon 9 rocket, known as "Block 5." This booster featured myriad changes to make it more capable of many flights, at a high cadence. For example, the grid fins that steered the rocket were upgraded to titanium, for durability. The landing legs could be retracted by the recovery crew, instead of needing to be removed. And most of the internal valves and other components were hardened with the knowledge gleaned from flight data.

This new Block 5 variant was a truly beastly rocket. While it bore superficial resemblance to the original Falcon 9, its guts were utterly transformed, having undergone eight years of ruthless assessment by thousands of engineers to shave mass and increase performance. The Block 5 version stood 230 feet tall, 50 feet higher than the original Falcon 9, and it more than doubled its lift capacity, to 50,000 pounds from 23,000 initially. And this new machine could land and launch again rapidly. The world had never seen its like.

"For those that know rockets, this is a ridiculously hard thing," Musk said on the eve of the Block 5 rocket's debut in May 2018. "It has taken us since, man, since 2002. Sixteen years of extreme effort and many, many iterations, and thousands of small but important development changes to get to where we think this is even possible."

The crazy hard work paid off. All the Block 5 variant has done is fly, and fly again, and fly again. Not even Musk and his smartest engineers know how far they'll be able to push this rocket. Some cores have now flown twenty missions to space, and they're still part of the active fleet.

## Emergency brakes now!

For all the effort put into optimizing the Falcon 9 for reuse, the Block 5 upgrade represented just one piece of the puzzle. Launch pads and ground systems to support rapid turnarounds were equally important. An essential part of this infrastructure is the transporter erector, or TE, which moves the rocket from the hangar to the launch pad and stands it up to launch.

In the aftermath of the AMOS-6 failure, efforts in Florida pivoted to rapidly completing the rebuild of Launch Complex 39A. Throughout the fall of 2016 teams worked outside on the massive TE that would haul not just Falcon 9s but eventually Falcon Heavy rockets. The launch pad sits atop a hill, about one quarter of a mile from where SpaceX constructed its hangar—a straight shot along some old railroad tracks.

After the 1-million-pound TE was completed in October, the launch pad team led by John Muratore sought to move it downhill to the hangar for the first time. For this job SpaceX had acquired, at great expense, two of the biggest available tug-tractors, the kind used for towing and push-back of large Boeing 747 and Airbus A340 aircraft. These would slowly guide the TE down the hill, and then push it back up with a Falcon attached. The rocket for the first launch from the rebuilt pad, the tenth cargo supply mission for NASA, awaited in the hangar.

As a group of SpaceXers looked on, including Muratore and Zach Dunn, then senior vice president of production and launch, the tugs started moving with frequent brake checks. On the level surface, at the top of the hill, everything worked fine. Similarly, at a slight gradient, the brakes held. But as the TE reached a steeper part of the hill, about a 5 percent gradient, the tug drivers began standing up, energetically applying as much force as they could to the tug brakes. It was no good. The tug brakes locked up and the TE bumped into them, driving them down the hill.

"I looked over at John, and he looked at me," Dunn said. "His eyes were as big as two saucers."

There was one last resort. As an emergency backup, engineer Chris Wallden had installed an independent braking system for the railway. The day before, during a test readiness review, Muratore had assigned the task of activating the emergency brake at his command, without hesitation, to a lead technician named Andy Clark. So as the tower rolled downhill, in his best commanding tone, Muratore shouted, "Emergency brakes now!" Clark ran to the side of the TE and threw a big manual lever to actuate the locomotive braking system.

"Just like in the movies, when you slam the brakes on a rolling train, they started squealing as the TE rolled down the hill," Dunn said. "And there was a split second there where it was not clear it would stop. And if it did not stop, this baby was going right down the hill, into the hangar, into the rocket, and right out into Kennedy Space Center."

But slowly, if barely, the TE slowed its descent. Amid ear-splitting squealing, it finally stopped. The disaster was averted.

SpaceX still needed a means of moving the TE to and from the launch pad. The tugs weren't going to work. Ultimately, they realized that the large winches being used by a salvage company to dismantle the old NASA launch tower could be repurposed to pull the TE up and down the hill. A few months later, on February 19, 2017, the CRS-10 mission launched from the rebuilt pad. It marked the first uncrewed launch from this site since 1973, when NASA put the Skylab space station into orbit.

As always, the goal was to move quickly. A consistent and valid criticism of SpaceX is that it always over-promised on launch numbers and under-delivered. By early 2017 the company was fifteen years old but had successfully launched just thirty rockets. That was a paltry average of two per year. Finally, however, SpaceX would start to deliver in Florida. To accomplish this, Lim and Muratore split their duties. Lim focused on refurbishing and reusing Falcon 9 rockets as quickly as possible, while Muratore focused on optimizing turnaround times.

During the spring of 2017, as Lim's launch team began reusing first stages, Muratore's pad team cut the turnaround time at Launch Complex 39A to two weeks. Still, they could do better. After a Bulgarian communications satellite lifted off on June 23, Muratore crunched the numbers. He believed that if everything went right, the pad could be ready for its next launch in just nine days. He called the pad team, mostly technicians, in for a meeting. Did they want to work all hours to support a July 2 launch? Or did they want a slightly easier flow, with a few days off for the Fourth of July holiday?

"I thought they were going to stand down," Muratore said. "But when I offered everyone a chance to move the launch to the other side of the holiday, the people turning the wrenches in the middle of the night said they wanted to go for it. Their view was that they were there to launch rockets, and that this was what it was going to take for the company to be successful."

SpaceX appealed to people who wanted to break records and weren't afraid to bust their asses. Musk and Shotwell had made it clear that, to become successful, SpaceX must launch frequently. Due to its low prices, SpaceX had acquired a large backlog of commercial satellite contracts, but those were worthless if the rocket could not get those satellites into space in a timely manner. Employees had stock options and a stake in its success.

So they did it. Nine days later SpaceX had another customer, satellite services provider Intelsat, on the pad. The countdown reached T-10 seconds when a flight computer aborted the attempt due to anomalous data from the flight guidance system. SpaceX found the rocket to be fine and reset the countdown for the next day, July 3. However, the countdown again automatically aborted at T-10 seconds. This time, a ground systems computer triggered the scrub.

Two last-second aborts at precisely the same time irritated Musk, then traveling in Europe. He cut his trip short and flew back to Florida for a debrief. "It was not the most pleasant meeting that I've been in," Muratore said. Musk wanted to know why the scrubs had happened and what could be done to make sure they did not happen again. "He was a little embarrassed that we had gone out there two days in a row, banging our heads into the wall," Muratore said.

After reviewing data on the Fourth of July, the launch team prepared for a third attempt. As usual, Muratore wore his "lucky" chrome shoes. This tradition dated to the previous November, when he had married his wife, Mary, at Cape Canaveral. The Air Force invited the couple to say their vows at Launch Complex 14, the historic pad where John Glenn reached orbit in 1962. Mary told Muratore his tuxedo was too plain and found a pair of patent leather shoes with a brilliant chrome finish.

After the ceremony, officiated by SpaceX engineer Trip Harriss, Gwynne Shotwell paid her respects. She had her own pre-launch tradition of writing "Scotland" on a sticky note and putting it in one of her shoes. This way, Shotwell can say she is "in Scotland" for a launch, where she was during the first successful liftoff of the Falcon 1. "These are your lucky shoes," she told Muratore. "Start wearing them for every launch."

He did, and finally on July 5, the Intelsat mission lifted off toward geostationary orbit. The year was only half over, and SpaceX had just launched its tenth rocket. Its previous best in a single year had been eight launches. Muratore would be wearing his lucky shoes a lot more frequently in the coming months.

The pad team continued fighting hard, but to reach Musk's aspirations of rapid reuse they needed to fight harder still. They had a difficult and dirty job, especially with the TE. This consisted of two parts, the transporter and the "strongback" that supports the vehicle up until the very last moment before liftoff, providing power and topping off propellant and other fluids. A launch is a bit like a birth, in that a rocket remains in the womb of the strongback until the final fractions of a second. The plumbing lines are actually called umbilicals. And there are many umbilicals, miles upon miles of electrical wiring and fuel lines and valves. By necessity, these lines are very close to the rocket and engulfed in fire at liftoff. After getting fried like a potato chip, all these components had to be inspected and often repaired between flights. This had been an especially bad problem at SLC-40.

"It would all catch fire a lot of the time," Muratore said. "All of the wiring was destroyed, and a lot of the valves would melt. The tubing was broken. So we had to go restore all of that after every launch. It was a huge mess."

As part of reconstructing the pad after the AMOS-6 conflagration, Musk wanted an upgrade. Originally Muratore's team proposed building a large and hardened steel transporter erector. But Musk was having none of it. "When we took the concept to Elon, he brutally shot it down as hideous and expensive," said Ryan Carlisle, a mechanical engineer who worked on its design. Soon after this takedown, Carlisle and two hydraulics engineers took respite at the Eureka! restaurant next to the Hawthorne headquarters. They started drinking Belgian beers, and after a few hours they had sketched a new concept called a "strongback throwback."

The original strongback at the launch site retracted only about 12.5 degrees away from the rocket before liftoff of the Falcon 9. Carlisle's solution pulled the strongback rapidly away from the rocket, and much farther, a full 45 degrees, to protect the umbilicals. In principle, this seemed simple. But in practice, it seemed an impossible task. At a quarter of a million pounds, the strongback had to move quickly. If it pulled back a second early, it would rip away the umbilical lines, stranding a fully fueled rocket with no connections. And if it were a second late, the strongback would just get toasted as usual. The kinds of hydraulic actuators needed for such a precise and powerful throwback did not exist in the United States, so Muratore's team sourced them from Bosch Rexroth, a European engineering firm.

Almost a decade after the original construction of SLC-40, SpaceX still adopted a scrappy mentality. The project had about a $50 million budget, but for Carlisle there was an upside to pinching pennies. It's how he met his wife, Gabrielle Inder, a SpaceX financial analyst who mapped out the project's budget.

"It's absolutely remarkable how little we spent building and rebuilding our launch pads at SpaceX compared to NASA and ULA," Carlisle said. "Literally over an order of magnitude cheaper."

All of the effort was worth it, because the strongback throwback worked, and with the addition of armored sections and an improved water deluge system, pad turnaround times continued their downward trajectory. The present record is two days and twenty hours, set in April 2024 at SLC-40.

SpaceX had to push the Range officials as well, particularly with the rocket's flight termination system. Under existing protocol, an Air Force official had responsibility for commanding a wayward rocket to blow itself up before it veered too far off course. SpaceX wanted to delegate this decision to a computer onboard the rocket. Although an autonomous flight termination system had been discussed for years, no company had ever designed, tested, or implemented one.

Prior to SpaceX's arrival, launch operations at Cape Canaveral and Kennedy Space Center had grown sleepy, especially with the end of the space shuttle. From 2000 to 2018, Florida rarely launched more than fifteen rockets a year, and sometimes there were fewer than ten flights. Over the years, the Range operators developed incredibly cumbersome safety requirements that took days to set up for each mission. To track a rocket's flight, operators use a sophisticated array of radars and telemetry. Everything was set up days ahead of launch, and once the destruct frequencies were dialed in, the Range was locked down. When rockets launched once a month, this was not a problem. But SpaceX wanted to go much faster.

For months ahead of a NASA cargo flight in early 2017, SpaceX engineers provided flight termination data to Range officials. Once again Howard Schindzielorz, who had assisted SpaceX with implementing the intent of regulations rather than the letter of the law under Susan Helms a decade earlier, stepped in. He supported SpaceX's efforts and pushed for the change. Wayne Monteith, the 45th Space Wing Commander who had greenlit their efforts to land rockets at the Cape, once again went along with a company asking him to put his commission on the line.

"My senior leadership was just not convinced it was the right thing to do," Monteith said. "But I had reviewed all of their analysis, and we'd flown it in shadow mode. To get permission I had to take responsibility for any failures. And I did, because I believed it was necessary to get the kind of cadence you're seeing today."

After Monteith signed off on the autonomous system, the cargo mission lifted off without incident. Eventually, the Air Force realized this was actually safer than the traditional way, as the onboard computer could react four seconds faster. This gave the rocket a little bit more time to correct its flight before it needed to trigger an explosive package. Within a couple of years, the Air Force embraced the technology, citing it as an innovative example of how it was preparing its launch ranges for the future.

As SpaceX flew more frequently, the culture at the Eastern Range started to change. When they were scarce, launches were an event. Mike McAleenan, the Air Force meteorologist, characterized the military's mobilization for a launch like D-Day, as if they were going to war. Officers would establish precise timelines for preparations and put all systems into a state of heightened alert. Finally, a day or two before liftoff, the Air Force and launch company would conduct a lengthy Launch Readiness Review meeting that lasted ninety minutes or longer.

SpaceX took some of the magic out of this. Those D-Day mobilizations? They're no longer necessary, as SpaceX and the Range have worked to automate preparations. And the momentous readiness meetings? They're perfunctory, and probably could be done over email. To the great consternation of space collectors, SpaceX no longer even produces a mission patch for every launch. As usual, the explanation falls back to airplanes. No one makes a patch every time one takes off.

"They've really pushed the Range to become more like an airport," McAleenan said. "And to a large part, they've succeeded."

He can think of only a single serious downside to the increased cadence. In the early days, SpaceX threw incredible parties after every launch, wild celebrations where the engineers and technicians released their pent-up stress and let loose. McAleenan was always invited, and for a time he would bring his shot ski. For the uninitiated, this is an old ski with three to six shot glasses attached in place of the ski bindings. The glasses are filled with liquor, and then several people hold the ski, slowly tipping it over and each downing a shot.

After a Falcon 9 launch of a telecommunications satellite for a Thailand-based company, Thaicom, McAleenan dutifully brought his shot ski. He lined up with Gwynne Shotwell, a "big dude from SpaceX," and two Thaicom executives dressed to the nines in suits. When the tall SpaceXer lifted the ski to drink, the sharp disparity in size caused the shots to be poured all over Shotwell and the shorter Thai officials.

After such a faux pas, future launch contracts were at risk. Word soon came down that, from now on, McAleenan's shot ski should stay home.

## Blue Origin email

Early in September 2018, Bob Smith sought to rally his employees. By then Smith had been chief executive of Blue Origin for a year, hired by Amazon founder Jeff Bezos to run his rocket company. Bezos first tried to poach Shotwell. After she declined, he eventually selected Smith, who came from Honeywell Aerospace, a fairly conventional company.

Smith had a big job ahead of him. Although Blue Origin was two years older than SpaceX, it lagged far behind in accomplishments. At the time Blue had not launched anything—not a satellite, not a spacecraft, not even something so insignificant as its logo, a feather—into orbit.

As he tapped out an email, Smith thought about an event Musk had staged the night before at SpaceX's headquarters in Hawthorne. A Japanese businessman, Yusaku Maezawa, revealed that he had signed up to fly a sortie around the Moon on SpaceX's Starship vehicle. "I'll tell you, it's done a lot to restore my faith in humanity," Musk said, seated in front of the end of a Falcon 9 rocket and its nine engines. "That somebody is willing to do this, take their money and help fund this new project that's risky, might not succeed, and is dangerous."

Maezawa was Starship's first paying customer, but Smith was not impressed. He viewed Starship as a distraction from SpaceX's core business, the Falcon 9 rocket. And so Smith emailed his company the following:

> **From**: Bob Smith
> **Sent**: Tuesday, September 18, 2018 7:30 AM
>
> Given the SpaceX announcement and our benchmarking of their best practices, there's a lot to be inspired by here and opportunities for us to catch up and surpass them as they get distracted by BFR and Starlink.
> -RHS-

The "BFR" mentioned is what Starship was then known as. In polite circles, it meant Big Falcon Rocket. Everywhere else, it simply meant Big Fucking Rocket, because that's what it was. Smith believed Musk's fixation on Starship and Starlink would give Blue Origin time to surpass SpaceX in the launch business. Bezos's own large rocket, named New Glenn, was due to make its debut in 2020.

Smith could not have been more wrong. In the first five years after Smith sent the email, SpaceX launched more than 175 additional orbital rockets, including Starship. Blue Origin has yet to launch any. Its New Glenn rocket should finally fly in 2024.

Smith also fundamentally misjudged the commitment Musk and SpaceX made to Starship and Starlink. This backfired badly when, in April 2021, SpaceX won the much-coveted contract to land humans on the Moon as part of NASA's Artemis Program to return astronauts there. The next humans to walk on the gray and dusty lunar surface almost certainly will do so by stepping off Starship. Bezos and Smith were incensed and sued NASA for a second chance. As for Starlink, by the end of 2023 it started to turn a profit for SpaceX.

Smith, who was finally fired in September 2023, is far from alone. The history of SpaceX and its competitors has rhymed over the last fifteen years since the boisterous birth of the Falcon 9 rocket on a cloudy night in Central Texas. First, these competitors gasp at Musk's ideas and explain why they are preposterous. Land a rocket at sea? *No way*. Refly a booster? *It's just not economical*. Build and launch the world's largest and most powerful rocket? *Not without the government*. Then these competitors fail to recognize reality until it is too late, before finally scrambling to emulate Musk.

To end this book, you might find it insightful to consider the rise of SpaceX, growing as it were from David into Goliath, from the perspective of its closest competitors. From these reflections it becomes clear how SpaceX has thoroughly transformed the global space industry.

## ULA faces a stunning defeat

Let us start with United Launch Alliance, SpaceX's earliest competitor for launch services in the United States. For more than a decade from its inception in 2006, the company co-owned by Boeing and Lockheed Martin launched virtually every NASA science mission of note, all of the big missions to Mars and beyond, and every national security payload, from spy satellites to GPS machines to the military's small space plane.

But then, in 2016, the Falcon 9 rocket tied United Launch Alliance's workhorse booster, the Atlas V, in total launches. That was the last year the two companies were close to one another. The next year, SpaceX launched eighteen rockets, more than doubling United Launch

Alliance's tally of seven Atlas and Delta boosters combined. The gulf has only widened since then.

By 2021 the "competition" between SpaceX and United Launch Alliance had increasingly become a farce. Once the dominant American rocket company that sought to squelch SpaceX, ULA launched a grand total of five rockets that year. By way of comparison, SpaceX launched five Falcon 9 rockets during the month of December alone.

One of the first stages that flew in December 2021 was a sooty booster that had made its debut nearly two years earlier, lofting the Crew Dragon Demo-1 mission into orbit. That rocket's latest flight was rather mundane, boosting yet another passel of Starlink satellites into space. However, in doing so, the core set a record, breaking Musk's goal of ten flights per booster with an eleventh mission. And the launch proved remarkable for another reason. During the lifespan of this well-worn booster, the Atlas fleet of United Launch Alliance had flown a grand total of eleven missions. A single Falcon 9 booster, therefore, matched the performance of eleven expendable Atlas rockets. Eleven Russian rocket engines lay on the bottom of the ocean. But at the surface, nine American engines stood atop a barge, ready to fly again.

The U.S. rocket wars were over. SpaceX had won.

Since then, SpaceX has kept beating the dead horse. Over one stretch, from the end of 2022 into the first half of 2023, SpaceX launched more than fifty rockets between United Launch Alliance flights. It has become difficult to remember that these two companies were once rivals, or that ULA's employees would drive up to the SpaceX fence, jeering.

For so long, SpaceX was disregarded and told what they could not do. Now, they're doing it, and their rivals aren't watching through the fence and laughing. They're looking up, in astonishment.

"We've come a long way, that's for sure," Gwynne Shotwell told me when I asked about disrespect. "We've heard, 'Oh, you're never gonna launch successfully.' We did Falcon 1. 'You're never gonna launch a real rocket.' We did Falcon 9. 'You're never gonna get Dragon to orbit. Dragon will never attach to station. Oh, you're never going to recover a rocket. Oh, you're never going to refly.' So it's kind of like, 'Screw you guys.'"

## Boeing's Starliner struggles while Dragon soars

SpaceX emerged triumphant over another major domestic competitor, Boeing, as well. The company that supposedly went for substance, rather than pizzazz, ended up with neither in the Commercial Crew race.

In December 2019, as Crew Dragon's parachute team raced through its final tests, Boeing flew Starliner for the first time. It did not go well. Due to a software problem shortly after launch, Starliner captured the wrong time from its Atlas V rocket. About ten minutes into its flight, Starliner's "mission elapsed time" was off by eleven hours. Flight controllers also briefly lost contact entirely with the vehicle. Due to these problems, NASA did not allow Starliner to approach the space station. Then, shortly before its return into Earth's atmosphere, a software mapping error nearly caused thrusters on Starliner's service module to fire in the opposite direction.

Following this debacle, NASA's chief of human spaceflight, Doug Loverro, took the extraordinary step of declaring Starliner's flight a "high visibility close call." This designation for NASA's human spaceflight program falls short of "loss of mission" but is still rather rare. It had been last used by NASA after a spacewalk in 2013 when water began to pool dangerously in the helmet of astronaut Luca Parmitano.

The next summer Crew Dragon launched Doug Hurley and Bob Behnken to the International Space Station. In November of the same year, the first operational Dragon mission, Crew-1, carried four astronauts to the station. And Dragon kept flying. In 2021, two more crew missions followed for NASA, as well as the first ever all-private orbital spaceflight. This Inspiration4 mission, spearheaded by billionaire Jared Isaacman and carrying a crew of four civilians, started to make good on SpaceX's promise to broaden access to space.

After more than two years of work on Starliner's software as well as sticky propulsion valves, Boeing finally flew Starliner on a redo of its first, uncrewed test flight in May 2022. This time Starliner did better, and the vehicle successfully docked to the space station.

By this time four years had passed since NASA named the crews for the first Dragon and Starliner missions. Back at that event in Houston, when Hurley had been so jubilant to fly with SpaceX, NASA had named three astronauts to Boeing's "Crew Flight Test." These were former space shuttle commander Chris Ferguson, shuttle pilot Eric Boe, and first-time flier Nicole Mann. During the intervening years, Ferguson pulled himself off the flight for family reasons and Boe did the same due to medical concerns. Mann, in her early forties, was seen as a rising star at NASA. The space agency was eager to get her flight experience so that she might be considered for Artemis flights to the Moon. The agency reassigned her to command the Crew-5 mission on Dragon, and she went into space in October 2022.

Boeing's Starliner woes continued. Heading into Memorial Day weekend 2023, mere weeks before the vehicle was finally due to make its crewed debut—with veteran NASA astronauts Suni Williams and Butch Wilmore—Boeing engineers found two more serious flaws. The first involved parachutes. Like SpaceX discovered years earlier, there was a problem with the lines that connected the spacecraft to the parachute's canopy. In the event of the failure of a single parachute, the lines between the spacecraft and its remaining parachutes could snap. The second issue involved glass cloth tape wrapped around wiring harnesses throughout the vehicle. These cables run everywhere, and there are hundreds of feet of them. The tape is intended to protect the wiring from nicks. However, it was discovered that under certain in-flight circumstances the tape is flammable.

These were shocking discoveries, especially so close to the flight. Neither NASA nor Boeing had good answers for why they had been found as astronauts were about to strap into Starliner. Questions emerged about the company's commitment to the program. Because it operates on a fixed-price contract, Boeing has reported losses of nearly $1 billion on Starliner. One of the ironies, of course, is that Boeing arguably saved the Commercial Crew program in 2009 by entering the competition.

## Dmitry Rogozin seems like he sucks

On the international stage, Russia was one of Musk's longest-standing adversaries. More than two decades ago, when he visited Russia to buy an old rocket for his plan to send a small terrarium to Mars, officials mocked Musk. Russia had launched the world's first satellite and the first human, and they had built the best rocket engines in the world. They had no use for a funny-talking weirdo from South Africa.

Today, Russia has an aging launch infrastructure. Its two main rockets, the Soyuz and Proton boosters, are based on technology developed in the 1960s. For a long time, this was good enough. Russia could boost its own military satellites, the Soyuz reliably got three people into space, and its launch prices were low enough to capture a decent share of the commercial satellite market.

The rise of SpaceX has exposed Russia's lack of planning for the future. At first, like a lot of SpaceX competitors, Russia downplayed the company's aspirations. As late as 2016, the Central Research Institute of Machine Building, which develops strategy for Russia's space corporation Roscosmos, dismissed the potential of rocket reuse. "The economic feasibility of reusable launch systems is not obvious," the institute concluded.

This report was published when Russia controlled about half of the global commercial launch industry with the Proton and Soyuz. But technical problems with Proton, as well as competition from SpaceX, began to erode this dominance. Toward the end of the decade, Russia's share fell to about 10 percent of the commercial satellites up for grabs. How did Russia's space chief respond to this? The plump and pugnacious Dmitry Rogozin tossed aside decades of Russian dominance in selling rockets as if it were an empty bottle of vodka.

"The share of launch vehicles is as small as 4 percent of the overall market of space services," Rogozin said in an interview with a Russian television station in 2018. "The 4 percent stake isn't worth the effort to try to elbow Musk and China aside. Payload manufacturing is where good money can be made."

But Russia would not make good money in payload manufacturing, or anything. After the Russian army invaded Ukraine in 2022, the international space community largely turned away from the Russian space program. The European Space Agency, which had bought Soyuz rockets for years, immediately stopped. And thanks to SpaceX, the United States could also break many of its relationships with Russia.

This would not have been possible even a few years earlier. For most of the 2010s, NASA relied entirely on Soyuz to carry its astronauts to the International Space Station. Crew Dragon broke that dependence. Also, the vast majority of military satellites flew on the Atlas V, United

Launch Alliance's mainstay, powered by a single, extremely powerful RD-180 engine manufactured in Russia. The Falcon 9 broke that dependence as well.

After Russia's invasion, the United States levied widespread sanctions, including on the aerospace industry. Rogozin's response was apoplectic. Among other threats he made, Rogozin said, "If you block cooperation with us, who will save the ISS from an uncontrolled deorbit and fall into the United States or Europe? The ISS does not fly over Russia so the risks are all yours. Are you ready for them?"

President Vladimir Putin eventually sacked Rogozin as the leader of Roscosmos, dispatching him to the front lines in Ukraine. Rogozin was seriously wounded a few months later, while dining at a cafe bombed in occupied Ukraine. He recovered, but it is unlikely that Russia's once-dominant space industry ever will.

## The steamroller arrives, at last

In 2014 a group of French launch industry officials visited SpaceX's factory in Hawthorne, California. Among them was Michel Eymard, the director of launch vehicles for the French space agency CNES. The urgency and scope of Falcon 9 operations impressed Eymard, and he returned home with detailed assessments for his government. He captured the essence of what he saw in a simple term, *rouleau compresseur*.

"We don't know when it will arrive, but the steamroller is on the horizon, and when it arrives it's going to be a real challenge for us," Eymard said in a remark that was later reported by Paris-based space journalist Peter B. de Selding.

With its early Ariane rockets, Europe practically invented the concept of commercial launch services. Prior to about 1990, if you had a satellite you wanted to get into space, you had to go to your nation's government. But starting with the Ariane 4 rocket, and later the Ariane 5, Arianespace would fly your satellite into space for the right price. A majority of the large, commercial satellites in geostationary space today, big birds that deliver telecom services such as DirecTV into homes, flew on Ariane rockets.

By 2014 some of the savvier observers, like Eymard, could see the coming storm. At that moment the continent faced a difficult choice. The Ariane 5 rocket was successful, but also expensive, costing tens of millions of dollars more than the Falcon 9. European officials, particularly from the countries of France, Germany, and Italy, met to decide how to respond. Should they seek an incremental upgrade of the existing Ariane rocket, with a focus on reducing its price? Or should they embrace the potential of reuse, and develop a revolutionary twenty-first century rocket?

Ultimately, with the Ariane 6, Europe chose the more conventional path. The rocket would be fully expendable and use solid rocket boosters like its predecessor. But with luck its price would be competitive with the Falcon 9 when it debuted in the year 2020. Perhaps this would save the continent from the impending steamroller?

Perhaps not. As often happens in rocket development, the Ariane 6 fell years behind schedule. This delay grew increasingly uncomfortable when the venerable Ariane 5 retired in the summer of 2023, and Europe's new rocket was not yet ready. This crisis drew forth the kind of remarks that were, previously, almost unheard of from SpaceX's competitors. With refreshing candor, the leader of the European Space Agency, an Austrian space scientist named Josef Aschbacher, acknowledged the reality of SpaceX's rise and its consequences for Europe.

"SpaceX has undeniably changed the launcher market paradigm as we know it," he wrote in a commentary. "With the dependable reliability of Falcon 9 and the captivating prospects of Starship, SpaceX continues to totally redefine the world's access to space, pushing the boundaries of possibility as they go along. Europe, on the other hand, finds itself today in an acute launcher crisis."

It was absolutely true, every word of it. But few rivals ever said these kinds of things out loud. To do so meant that one had ignored the threat of SpaceX early on and delayed reacting in a meaningful way for too long. By the early 2020s, however, these facts were ineluctable.

Elon Musk and his rocket company now stand alone, atop the hierarchy of spaceflight. SpaceX launched nearly 100 rockets in 2023, about the same total as the rest of the world combined, including China and its dozens of launch startups. It may reach 150 launches in 2024. SpaceX operates more satellites than anyone, anywhere, by a factor of ten. Crew Dragon carries as many humans into orbit as the human spaceflight programs in Russia and China combined. Then there is Starship, the largest and most powerful, and potentially most disruptive, rocket ever built. SpaceX has lapped the world in spaceflight, and with Starship, the steamroller is coming around for another pass.

A few weeks after remarking on the European launch crisis, Aschbacher traveled to the United States in early July 2023. He was eager to see the safe launch of the Euclid space telescope, a fantastic new science instrument that will measure the rate at which our universe is ripping itself apart. It had taken a decade and $1.5 billion to construct the telescope, making it one of the costliest instruments of its kind ever built by the European Space Agency. Originally, Euclid was due to launch on a Soyuz rocket from a European spaceport. Then the war happened, and the Ariane 6 rocket suffered more delays.

The world turns. Dynasties fall, upstarts rise. David becomes Goliath.

In Florida, due to the sultry summertime air, Aschbacher traded his customary bright blue suit for slacks and a polo shirt. He mingled among other VIPs on the balcony of a NASA building with a splendid view of the launch site several miles away. Peering into the distance, Aschbacher could just make out a black-and-white rocket surrounded by four tall lighting towers. It started to vent puffy clouds of oxygen.

Far above the Falcon 9, destiny beckoned.

# EPILOGUE

Four days after astronauts Bob Behnken and Doug Hurley triumphantly splashed into the Gulf of Mexico in August 2020, Elon Musk hit send on a companywide email. Though Dragon had barely been lifted from the water, it was time to pivot. "Please consider the top SpaceX priority to be Starship," Musk wrote. The company needed to "dramatically and immediately" accelerate progress. He urged employees to seriously consider relocating to South Texas to work more directly on Starship.

Here was Musk distilled to his essence. During the inaugural launch of Falcon 1, he focused on buying aluminum for the Falcon 5. After just the second successful launch of the small rocket, Musk canceled Falcon 1 to go all-in on Falcon 9. On the eve of this larger rocket's debut, he sat down on the pad, fretting about how to increase launch cadence. When I met Musk in 2018, a day before the first Falcon Heavy launch, he looked ahead to Starship. And in the summer of 2020, after delivering a huge win for NASA and the United States with Dragon, Musk asked his team to pivot yet again.

This is how SpaceX became a steamroller. As Tom Mueller said, SpaceX development programs were always way slower than Musk wanted and way faster than anyone had gone before. And when SpaceX achieves some crazy feat or reaches a dizzying new height? Musk moves the goal posts. The steamroller must keep accelerating to reach Mars.

But will it? That's the question I kept coming back to as I wrote *Reentry*. Can the SpaceX steamroller be stopped? I worry the answer might be yes. I want to close this book with some meditations on why. I cannot know the future, but I can share a few thoughts about where things may be headed. This is solely my perspective after talking to hundreds of SpaceXers and key industry sources, and spending thousands of hours thinking and writing about this topic.

For the sake of transparency, I will first tell a little bit about myself. I've had a lifelong interest in space. A fascination with the stars led me to study astronomy in college, which involves a lot of physics and calculus and precious little stargazing. None, actually. I went into a career writing about science because four semesters of calculus were enough math for me. After more than a decade as a middling science writer at the *Houston Chronicle*, I discovered SpaceX. Musk's compelling vision for reusable launch and ultimately settling other worlds in the solar system captivated me. *Hell yes, we should do this*, I thought. This is the future I wanted. And this guy had a plan. I started writing about space with passion and developed a special interest in the commercial space revolution. A decade later I'm privileged to have written my second book on SpaceX and to possess a voice in the space industry.

I'm going to use that voice here because I care deeply about the success of our human spaceflight endeavors, particularly NASA's Artemis Program to explore the Moon and beyond. SpaceX is the lynchpin for Artemis becoming more than a pale echo of the Apollo Program. NASA can only afford to go back to the Moon because it is saving billions of dollars by using Starship. And only with the full reusability of Starship is truly sustainable deep space exploration possible. Artemis will go as far as Starship goes. And if Artemis is a success, it opens the pathway to Mars for NASA and SpaceX, and the red planet's eventual settlement. The future is right there for the taking, and we dare not miss this chance.

But I am worried about Elon. There is exactly one person who can stop the steamroller, and that is the same person who set it into motion and still has his foot pressed down to the floor, pedal to the metal. During the period in which I wrote *Reentry*, Musk purchased Twitter (renamed X) and in a matter of months destroyed much of its value. As I watched the immolation of $44 billion spent to acquire the social network, I struggled to square how the person who led SpaceX to such heights could subsequently make these seemingly poor decisions, driving users away from Twitter, flirting with antisemitism, and amplifying a lot of other bullshit that has infected American culture in the twenty-first century. More than once, I found myself asking, "What the hell are you doing, Elon?"

I remember vividly, during the Guadalajara speech in 2016, Musk speaking passionately about his Mars vision. At one point he said: "The main reason I'm personally accumulating assets is in order to fund this. So I really don't have any other motivation for personally accumulating assets except to be able to make the biggest contribution I can to making life multiplanetary." This line earned him a rousing round of applause. Eight years later, it rings false. Buying Twitter cost Musk a chunk of his personal fortune and an even bigger slab of credibility with investors. It is difficult to see how X is contributing to making life multiplanetary.

As SpaceX rose from obscurity to dominance over the last fifteen years, Musk's presence on the global stage similarly expanded. His Twitter acquisition has only further magnified his fame. And he seems to crave the newfound attention. With a carnival barker's desire to stand in the center of the spotlight, Musk injects himself into the debates of the day. Musk did not purchase Twitter to make money. He purchased it to amplify his views and those with whom he agrees. This is his right. But there may well be costs for SpaceX.

This is what I worry about.

## Sizing up the competitors

No one will topple SpaceX anytime soon. Thanks to Musk's relentless drive, the company has no peers in spaceflight anywhere. But there are some prospective challengers. Blue Origin has the most potential, and Amazon founder Jeff Bezos seems willing to continue pumping billions of dollars into the company every year for little financial return. I'll give Bezos this: Blue Origin is world-class at building facilities. But so far, they have struggled with building big rockets. Hopefully that will change with the New Glenn rocket. Yet here's the thing. At the end of 2023 the space company had 11,000 employees, about the same as SpaceX, with comparatively few accomplishments.

"They are the world's largest single-donor nonprofit," a person who worked at both SpaceX and Blue Origin told me. "There is zero incentive to operate like SpaceX. Like, zero. I know I'm going to get fed no matter what. The doors will never close. They start off with a huge disadvantage there. When you have a funding model like that, you attract the dreamer who says, 'Where can I go to work on something crazy, and make a good salary at forty hours a week? Where can you work on a space elevator and get paid? Who are the types of people who want to work on space elevators?' So there's a selection bias. You need to take extraordinary measures to account for that, and they don't have them."

Blue Origin seeks to compete with SpaceX not just on launch, but in other areas as well. The company is building a fully reusable lunar lander for NASA's Artemis Program. (It's years behind Starship.) That effort is being led by a familiar face, John Couluris, who led the C2 mission for SpaceX in 2012. I hope he crushes it. Bezos also has a Starlink competitor called Project Kuiper. I have less confidence in that initiative as it is led by Rajeev Badyal, whom Musk fired as director of the Starlink program for moving too cautiously. Embarrassingly, in late 2023, Bezos resorted to buying launches on the Falcon 9 to get Kuiper satellites into orbit because Blue Origin's rocket was so long delayed.

Having a robust competitor to SpaceX would be a real boon to the U.S. space industry, but we have been waiting on Blue Origin to step up for a decade. I am rooting for them. Whatever you may think about Bezos, his passion for space and belief in off-world settlement is genuine. I just have not seen enough to inspire confidence they're going to get there anytime soon.

In the United States there are SpaceX alternatives more efficiently run, most notably Rocket Lab. The company's founder and chief executive, Peter Beck, is a pretty amazing dude. If you read Ashlee Vance's book, *When the Heavens Went on Sale* (Ecco, 2023), you can't help but be impressed by how far Beck has come, starting from nothing and nowhere. However, Rocket Lab has substantially fewer resources to hand than SpaceX or Blue Origin. I eagerly anticipate Rocket Lab's new medium-lift launch vehicle with a reusable first stage, but it is more than a decade behind the Falcon 9. Frankly, I believe Bezos should buy Rocket Lab and empower Beck. That has the best chance of molding Blue Origin into a true SpaceX competitor.

There are other promising new space companies in the United States, including Relativity Space and Stoke Space, with innovative approaches to reusable launch. But these companies are essentially where SpaceX was during its Falcon 1 days. If you have learned anything from reading this book, I hope it's how long and hard the struggle is to get to a robust, reusable rocket like the Falcon 9.

With respect to space programs in Europe, India, and Japan, the only other competitor with the breadth of ambition and resources to challenge SpaceX for supremacy in global spaceflight is the combined efforts of the Chinese government and its quasi-commercial industry. China's space program rapidly ascended in the last decade, flying increasingly complex interplanetary robotic missions and constructing a space station in low-Earth orbit. And perhaps more than any other competitor, Chinese rocket scientists have taken note of SpaceX's achievements and sought to emulate them. This is no sure thing, however, given the fiscal uncertainties in China and a form of government that does not embrace the kind of free-market capitalism that has allowed Musk and SpaceX to flourish.

For all of these reasons, I believe that if SpaceX is to be superseded in the next decade, it will not be due to someone else's rise, but because of SpaceX's fall.

## The trouble within

Do you want to know one of the things that amazes me most about SpaceX?

After two decades, the company has not lost its founder's mentality. Most startups form as insurgents in their industry, with a bold mission to serve customers in new and better ways. SpaceX did this by offering fixed-price contracts that gave NASA and other customers a better deal, delivering superior products faster at a better price. Crew Dragon is the most prominent example but not the only one. As most companies grow, they lose the founder's mentality, their hunger, their insurgency. But not SpaceX. They are still charging headlong into the future, hard and fast and sometimes recklessly, toward the next disruptive thing.

This is unquestionably because of Musk. Even today, when he steps away from SpaceX for an extended period of time to focus on Twitter or

Tesla or Neuralink or AI or whatever, the work slows down. He can be a distraction, certainly, but he also remains the company's essential and invigorative force. Musk accelerates progress by his presence and willingness to take responsibility. When he makes a consequential decision, he accepts the risk. In most other companies or government agencies, no senior manager wants to make difficult calls and be saddled with the consequences of failure. So they defer, studying a problem more or going through endless reviews and tests. Not Musk. He presses the "go" button. Sometimes the decision backfires spectacularly, but his willingness to accept the responsibility keeps things moving.

And when things don't work, there is no sunk cost fallacy. Even today, as part of the SpaceX employee performance review, one of the criteria is "responds positively to rapid changes in strategic direction." Musk's autocratic control of SpaceX, and willingness to pivot, allows employees to drop unnecessary burdens and move quickly in a new direction. His late-night decision to ditch the Falcon 5 in favor of a rocket design with nine Merlin engines offers but one of many, many examples of Musk identifying and then following the shortest path into the future.

Musk is at his best in uncharted territory. When he's blazing a trail, there is no one better in the world at looking around the corner for what could and should come next. For his employees, Musk removes barriers and constraints, encouraging them to expand their minds for what is possible, to free their thinking and search for unbounded solutions to engineering problems. It has worked astoundingly well at SpaceX.

Just as great individuals have great strengths, however, they have great weaknesses. Some of these failings have been on display in his running of Twitter/X: the vanity, the vindictiveness, and the double-talk about freedom of speech while curbing views he does not like. Social media is not rocket science, and Musk does not seem to be very good at running a company that does it. Moreover, Musk's purchase of Twitter has deepened his addiction to the service and, for complicated reasons, pushed him further from a Libertarian outlook into far-right conservatism and conspiracy theory mongering. Musk is absolutely entitled to his political opinions, but he is not required to share them or let them guide his decision-making at Twitter. He has alienated a lot of people.

As I write this, SpaceX has largely flown above the fray so far. Bill Nelson, a former Democratic senator who is NASA's administrator, has never particularly trusted or liked Musk. But he recognizes the value SpaceX brings to the space agency and its importance for the future. When I asked Nelson about Musk's behavior in the spring of 2023, he changed the subject to talk about Gwynne Shotwell and how she has his full confidence. Her presence is important for the company's military contacts as well. But Shotwell is now in her sixties and has been looking to retire. What happens then?

There is no clear successor. The technical leader at SpaceX is obvious: it's Mark Juncosa. He is not without flaws, but he is a charismatic and technically brilliant manager. He also knows how to work with Musk and talk him out of some of his crazier ideas. Any time Musk has needed a big problem solved, Juncosa stepped up. But he is not a replacement for Shotwell. He lacks the velvet gloves. As one SpaceXer told me, "We can't let Mark around customers." Too many rough and wild edges.

Looking ahead, my concerns are threefold. One, Musk could do something so egregious that the U.S. government stops doing business with his companies. This seems highly unlikely, at least in the near term. The Falcon rockets, Dragon spacecraft, Starlink satellite constellation, and Starship vehicle are essential to major U.S. civil and military initiatives. But decision-makers and political leaders are watching Musk and his actions, and they are wary of his immense power to shape global events like the war in Ukraine and the Israel-Gaza conflict.

There are plenty in Congress who would gladly pounce on Musk if given an opening.

My second concern is what happens if Shotwell retires. She engenders confidence in NASA, at the U.S. Department of Defense, and with customers. If there is not a similar replacement to counterbalance Musk, the rails could come off.

Finally, SpaceX could lose Musk. If he leaves, the company risks drifting from its founder's mentality. There are plenty of historical analogs. The original versions of Boeing, Lockheed, and Martin Marietta, not to mention the oil and rail baron conglomerates, all had eccentric leaders. Capitalism and corporate boards and government oversight eventually reined them in. So far Musk and SpaceX have resisted this, remaining a disruptive force. It is important for SpaceX and the Mars vision that this continues. The Falcon 9 is the most reliable, advanced, and cost-effective rocket in the world. And yet Musk is pushing his engineering teams every day to obsolete it as rapidly as possible with Starship. An established company does not do that. But a disruptive, innovative one does.

To remain on a glidepath for success, therefore, Musk faces huge challenges in the years ahead. He must remain an agent of change and disruption, while not becoming an agent of chaos and distraction. He must avoid the path of Howard Hughes, an American aerospace engineer who half a century ago descended from magnificent entrepreneurship into eccentric madness. In the years ahead, SpaceX may need to raise billions of dollars more in private capital, which will require a Musk whom investors can trust and be inspired by. NASA and spaceflight regulators will want the reliable partner they have had for nearly two decades. As part of this, Musk must also find a capable leader after Shotwell leaves to help steady the steamroller.

That's a hell of a lot, and I ardently hope Elon succeeds.

He has inspired me and countless others with the achievements of SpaceX and the brilliant potential of its future. He has made space interesting and fun and full of promise. He has disrupted the old-world order of spaceflight very much for the better, replacing stodgy and stale with daring and dynamic. What once seemed impossible no longer does. Humans belong among the stars. The barriers are falling away. The future is unknowable, but tantalizing.

I don't want this ride to end. It's just starting to get good.

—Eric Berger

# ACKNOWLEDGMENTS

This was a difficult book to write. During this narrative's timeline, from 2008 to 2023, more than ten thousand people were employed by SpaceX, to say nothing of those in the space industry and government who worked alongside them. It is a vast story, with many threads. The challenge came in trying to weave all of it together into a coherent narrative that captured the essence of what happened without overburdening the storytelling.

That's a long way of saying that I feel bad for the people I did not talk to and whose SpaceX stories are not captured in these pages. Every time I interviewed someone for *Reentry* they would suggest a handful of other people I should talk to for more details, a first-person account, or other insights. I had to draw the line somewhere or *Reentry* would never have been written.

So I want to recognize everyone who toiled away over the last fifteen years to make the magic described within these pages happen. I know you worked hard and long and passionately. You made emotional and physical sacrifices. And in the process, you brought new and wonderful technologies into this world, changing the staid old ways of spaceflight forever. You have made a new and better future for humanity. My first acknowledgment is to you, the doers.

I am grateful to the people who spoke with me for this book. If you read *Liftoff*, you will see some old friends in these pages, including Tim Buzza, Hans Koengismann, Tom Mueller, and Zach Dunn. They spoke at length with me again for this narrative, providing invaluable details, fact-checking, and recommendations for whom to interview. There were a whole host of new people to speak with, who came to SpaceX late during the Falcon 1 era or after and made their indelible mark on the Falcon 9 and Dragon programs. I thank them all.

For this book I had far less access to Elon Musk and current employees at SpaceX. Musk said he was eager to participate, but every time we got close to an interview it would be canceled. I did a lot of the work on *Reentry* during the time when he was talking to Walter Isaacson for his biography, and in the messy process of buying and transforming Twitter. Both of those were likely factors. Fortunately, I experienced the years chronicled in this book as a space journalist, with good access, and have spoken with Musk many times while these events were unfolding. Moreover, a majority of the key players during the years of this book had left SpaceX by the time I started writing. The end result is that *Reentry* is not so much an Elon book, but a book about his company and the engineers who made it such a fantastical success.

The U.S. cover of *Reentry* looks fabulous. I love the symmetry between the U.S. covers of *Liftoff* and this book, with a Falcon rocket and text rising on *Liftoff*, and a Falcon rocket and text falling on *Reentry*. I want to thank Sarah Avinger and her talented team at BenBella for the design, and uber-talented space photographer John Kraus for the gritty landing image. John is a good friend and has done much to broaden the popularity of spaceflight through his gorgeous imagery.

I also want to thank my agent, Jeff Shreve, for his help in my journey to becoming a successful author, and for his support as I go forward. For this book I have had two great editors at BenBella, Claire Schulz and Rick Chillot. Both were diligent in making sure that I stuck to the narrative and did not get too far bogged down in policy or other editorial alleyways. My goal with *Liftoff* and *Reentry* has been to bring readers inside the story and place them alongside the people building and launching rockets.

In addition to writing books, I have two day jobs, and *Reentry* took me away from both at times. I work for a wonderful editor-in-chief of Ars Technica, at Conde Nast, Ken Fisher. Ken is super patient and supportive. Early during my tenure at Ars I recall Ken saying in a staff meeting, "I'm the best boss you'll ever work for." It sounded like a hollow boast then, but it rings true now. He is the best. I work alongside some great people at Ars, including Eric Bangeman, Lee Hutchinson, John Timmer, and Stephen Clark. I also have a Houston weather website, Space City Weather, which I operate with another meteorologist, Matt Lanza. A tremendous partner, Matt was always willing to cover a forecast when I needed to travel or buckle down on something to do with this book.

Finally, there is my family. My dad gave me a love of writing, and my mom was a stickler for accuracy. In these pages I have attempted to tell the story of SpaceX as accurately as possible. I want to apologize to my wife, Amanda, and two daughters, Analei and Lily, for going missing for large chunks of 2023. There were several months where I started work at about 6 AM and, in between Ars Technica, forecasting, and writing, did not emerge from my home office until 10 or 11 at night. For a while there, I felt a bit like a SpaceXer on deadline. It allowed me to empathize with them, and their families, a little. Anyway, I'm sorry to my own family for the neglect, and I deeply appreciate your patience. I love you all so much. I hope I've made you proud.

# INDEX

# ABOUT THE AUTHOR

Photo Credit: Amy Carson Photography

**ERIC BERGER** is the senior space editor at Ars Technica, covering everything from private space to NASA policy, and author of the book *Liftoff* (William Morrow, 2021), about the rise of SpaceX. Eric has an astronomy degree from the University of Texas and a master's in journalism from the University of Missouri. He previously worked at the *Houston Chronicle* for seventeen years, where the paper was a Pulitzer Prize finalist in 2009 for his coverage of Hurricane Ike. A certified meteorologist, Eric founded Space City Weather, co-founded The Eyewall, and lives in Houston.